Intermediate 1
CHEMISTRY

Norman Conquest

Hodder Gibson

A MEMBER OF THE HODDER HEADLINE GROUP

The Publishers would like to thank the following for permission to reproduce copyright material:
Photo credits: Andrew Lambert Photography/Science Photo Library; Andrew Maroney; Andrew Mcclenaghan/Science Photo Library; Arnold Fisher/Science Photo Library; Bsip Taulin/Science Photo Library; Cc Studio/Science Photo Library; Charles D. Winters/Science Photo Library; Cordelia Molloy/Science Photo Library; Damien Lovegrove/Science Photo Library; David Taylor/Science Photo Library; Dr Jeremy Burgess/Science Photo Library; Erika Craddock/Science Photo Library; Firbimatic Uk; Francoise Sauze/Science Photo Library; Gusto/Science Photo Library; Jack K. Clark/Agstock/Science Photo, Library; James L. Amos/Corbis; Jeremy Walker/Science Photo Library; Kazuyoshi Nomachi/Science Photo Library; Lauren Shear/Science Photo Library; Martyn F. Chillmaid/Science Photo Library; Maximilian Stock Ltd/Science Photo Library; P. Dumas / Eurelios/Science Photo Library; Pascal Goetgheluck/Science Photo Library; Peter Bowater/Science Photo Library; Philippe Psaila/Science Photo Library; Planetary Visions Ltd/Science Photo Library; Robert Brook/Science Photo Library; Simon Fraser/Science Photo Library; Tony Craddock/Science Photo Library; © Adam Woolfitt/Corbis; © Adrian Carroll;Eye Ubiquitous/Corbis; © Amos Nachoum/Corbis; © Don Mason/Corbis; © Gary Trotter; Eye Ubiquitous/Corbis; © George Mccarthy/Corbis; © Philip Gould/Corbis; page 184 (bottom) © Cephas Picture Library/Alamy.

Acknowledgements
The author wishes to thank the following individuals for their help at various stages in the production of this book.
Robin Nicolson, Allen Shepherd and Maria d'Arcy of Perth Grammar School, also Margaret Robson of Heriot Watt University.
   Norman Conquest, 2004.

Although every effort has been made to ensure that website addresses are correct at time of going to press, Hodder Gibson cannot be held responsible for the content of any website mentioned in this book. It is sometimes possible to find a relocated web page by typing in the address of the home page for a website in the URL window of your browser.

SQA Examination questions are reproduced by permission of the Scottish Qualifications Authority, Dalkeith.

The cover photo shows fireworks exploding over Edinburgh Castle and Princes Street

While every effort has been made to check the instructions of practical work in this book, it is still the duty and legal obligation of schools to carry out their own risk assessments.

Orders: please contact Bookpoint Ltd, 130 Milton Park, Abingdon, Oxon OX14 4SB. Telephone: (44) 01235 827720. Fax: (44) 01235 400454. Lines are open from 9.00 - 5.00, Monday to Saturday, with a 24-hour message answering service. Visit our website at www.hoddereducation.co.uk. Hodder Gibson can be contacted direct on: Tel: 0141 848 1609; Fax: 0141 889 6315; email: hoddergibson@hodder.co.uk

© Norman Conquest 2004
First published in 2004 by
Hodder Gibson, an imprint of Hodder Education, part of Hachette Livre UK
2a Christie Street
Paisley PA1 1NB

Impression number   10 9 8 7 6 5 4
Year   2010 2009 2008

Cover photo provided by Stone/Getty Images (297359-003)
Designed and typeset in Utopia 11pt by Hardlines, Charlbury, Oxford
Printed in Dubai for Hodder Gibson, 2a Christie Street,
Paisley PA1 1NB.

A catalogue record for this title is available from the British Library

ISBN 13: 978 0 340 81699 8

To the Student

This book covers the Access 3 and Intermediate 1 Chemistry courses of the Scottish Qualifications Authority. The shaded sections are relevant only to the Intermediate 1 course.

There are short questions throughout the text to check your understanding of what you have read. Access 3 subsection tests and Intermediate 1 end of unit tests are provided to give you practice for the National Qualification tests. A selection of questions from past NQ examination papers is also provided.

Prescribed Practical Activities have been described at appropriate points in the text. Sufficient detail has been provided to satisfy the requirements of the external examination.

I hope that you find this book useful.

Norman Conquest 2004

Contents

Everyday Chemistry

Chemistry in Action

The topics covered in this unit are
Substances
Chemical reactions
Bonding
Acids and alkalis

1 Substances

Figure 1.1 *Everything on the Earth is made from about one hundred elements*

Elements

Everything that you see around you is made from about one hundred **elements** (see figure 1.1). Elements are the basic building blocks of the whole world. They are the simplest kind of substance and cannot be broken down into anything simpler. Hydrogen and oxygen are examples of elements.

All the known elements are given a special number called the **atomic number** and listed in the **Periodic Table**. There may be copies of this displayed on the walls of your school laboratory. You will have seen that, as well as a name, each element has been given a **symbol**.

In most cases the symbol for an element is the first letter or the first two letters of the element's name. For example, H is the symbol for hydrogen and Al is the symbol for aluminium.

In some cases the symbol for an element is based on its Latin name. In the past, most water pipes in houses were made from the element lead. The symbol for lead is Pb, which comes from the Latin word '*plumbum*', meaning lead. So you can see how plumbers originally got their name.

More recently discovered elements have been named after famous scientists. One has been named after the famous Russian chemistry teacher, Dimitri Mendeleev (see figure 1.2), who invented the Periodic Table.

Figure 1.2 *Dimitri Mendeleev, the inventor of the Periodic Table. (He cut his hair once a year, in spring, as the weather began to get warmer.)*

Section Questions

1 Use the Periodic Table on page 4 to find the symbols for:
   a) iodine,
   b) argon,
   c) magnesium,
   d) tin.

2 Write down the elements that were named after the following scientists:
   a) Ernest **Rutherford**,
   b) Niels **Bohr**,
   c) Marie **Curie**,
   d) Alfred **Nobel**.

   Use the same Periodic Table to help you. Start looking from element number 93 onwards.

Classifying the elements

If you have a lot of music on CDs, you may have found it useful to group them together as 'rock', 'soul', 'country and western' and so on. With more than one hundred known elements, it is also useful to put them into different groups. We call this classifying the elements.

a) Are the elements solids, liquids or gases?

This gives us three groups of elements. Most of the elements are solids at room temperature but a few are gases and two, mercury and bromine, are liquids. The gases include hydrogen, oxygen, nitrogen, fluorine and chlorine. There is also a family of very unreactive gaseous elements in the last column (column 0) of the Periodic Table.

Periodic Table of the Elements

Atomic number	
Name of element	
Symbol	

* Elements below the dark line are metals

Alkali Metals

Halogens

Noble Gases

Transition Metals

1 Hydrogen H																	2 Helium He
3 Lithium Li	4 Berllium Be											5 Boron B	6 Carbon C	7 Nitrogen N	8 Oxygen O	9 Fluorine F	10 Neon Ne
11 Sodium Na	12 Magnesium Mg											13 Aluminium Al	14 Silicon Si	15 Phosphorus P	16 Sulphur S	17 Chlorine Cl	18 Argon Ar
19 Potassium K	20 Calcium Ca	21 Scandium Sc	22 Titanium Ti	23 Vanadium V	24 Chromium Cr	25 Maganese Mn	26 Iron Fe	27 Cobalt Co	28 Nickel Ni	29 Copper Cu	30 Zinc Zn	31 Gallium Ga	32 Germanium Ge	33 Arsenic As	34 Selenium Se	35 Bromine Br	36 Krypton Kr
37 Rubidium Rb	38 Strontium Sr	39 Yttrium Y	40 Zirconium Zr	41 Niobium Nb	42 Molybdenum Mo	43 Technetium Tc	44 Ruthenium Ru	45 Rhodium Rh	46 Palladium Pd	47 Silver Ag	48 Cadmium Cd	49 Indium In	50 Tin Sn	51 Antimony Sb	52 Tellurium Te	53 Iodine I	54 Xenon Xe
55 Caesium Cs	56 Barium Ba	57 Lanthanum La	72 Hafnium Hf	73 Tantalum Ta	74 Tungsten W	75 Rhenium Re	76 Osmium Os	77 Iridium Ir	78 Platinum Pt	79 Gold Au	80 Mercury Hg	81 Thallium Tl	82 Lead Pb	83 Bismuth Bi	84 Polonium Po	85 Astatine At	86 Radon Rn
87 Francium Fr	88 Radium Ra	89 Actinium Ac	104 Rutherfordium Rf	105 Dubnium Db	106 Seaborgium Sg	107 Bohrium Bh	108 Hassium Hs	109 Meitnerium Mt									

58-71 ●

90-103 ■

58 Cerium Ce	59 Praseodymium Pr	60 Neodymium Nd	61 Promethium Pm	62 Samarium Sm	63 Europium Eu	64 Gadolinium Gd	65 Terbium Tb	66 Dysprosium Dy	67 Holmium Ho	68 Erbium Er	69 Thulium Tm	70 Ytterbium Yb	71 Lutetium Lu
90 Thorium Th	91 Protactinium Pa	92 Uranium U	93 Neptunium Np	94 Plutonium Pu	95 Americium Am	96 Curium Cm	97 Berkelium Bk	98 Californium Cf	99 Einsteinium Es	100 Fermium Fm	101 Mendelevium Md	102 Nobelium No	103 Lawrencium Lr

● ■

b) Are the elements metals or non-metals?

Some Periodic Tables have a dark zig-zag line running through them. This is to divide the **metal** elements from the **non-metal** elements. Metal elements are shiny and conduct electricity. Iron, copper, silver and gold are examples of metal elements. A typical non-metal element is not shiny and does not conduct electricity. Oxygen and sulphur are examples of non-metal elements. By looking at the Periodic Table on page 4 you will see that there are many more metal elements than non-metal ones.

Section Questions

3 By using the Periodic Table on page 4, state whether the following are metals or non-metals:
a) scandium,
b) arsenic,
c) silicon,
d) beryllium.

4 State the names of the two elements that are liquids at room temperature and classify each as either a metal or a non-metal.

Dates of discovery

Some elements, including gold, silver and copper, have been known since prehistoric times. Others have been discovered more recently, such as calcium in 1808 and aluminium in 1827. New elements are still being discovered, and it is these latest ones that are being named after famous scientists. For example, einsteinium was discovered in 1953 and is named after Albert Einstein.

Section Questions

5 Use the 'Date of discovery' table on page 2 of the data booklet to find the dates of discovery for the following elements:
a) hydrogen,
b) neon,
c) platinum,
d) americium.

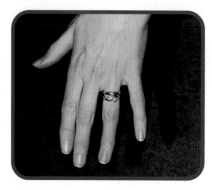

Figure 1.3 *Can you spot the carbon and gold in this photograph?*

Everyday uses for elements

Many elements have everyday uses. Some you will have heard of, others will be less familiar.

Some elements you can actually wear! **Gold** is used to make rings, necklaces and earrings (see figure 1.3). **Silver** can be made into bracelets. Did you know that diamonds are simply made of **carbon**?

There are other elements in homes apart from those in jewellery. Water pipes and hot water tanks are often made of **copper**. Copper is also used for electrical wiring.

What gas would you use in a packet of crisps? Air is no good because the crisps soon go bad. Nitrogen is better, but **argon** is best for keeping crisps tasting good for longer (see figure 1.4).

Huge amounts of **iron** and **aluminium** are made each year. Can you think what these metal elements are used for? A lot of aluminium is used to make aeroplanes. This is because it is light and strong. Iron is stronger than aluminium, but also heavier. A lot of iron is turned into steel for make cars, lorries, ships and bridges (see figure 1.5).

Figure 1.4 *This crisp packet contains the very unreactive gas argon*

Figure 1.5 *The Forth Rail Bridge contains thousands of tonnes of iron*

More about the Periodic Table

When Dimitri Mendeleev drew up the Periodic Table, he did so by putting elements that had similar chemical properties in the same columns.

Column 1 – reactive metals

| Li lithium |
| Na sodium |
| K potassium |
| Rb rubidium |
| Cs caesium |
| Fr francium |

All of the metals in column 1 are silvery but soon become dull, as they corrode quickly in air. They also react quickly with water (see figure 1.6), giving off hydrogen gas and producing alkaline solutions. Because of this, they are called the **alkali metals**.

Figure 1.6 *Sodium reacting with water*

Column 7 – reactive non-metals

| F fluorine |
| Cl chlorine |
| Br bromine |
| I iodine |
| At astatine |

Column 7 contains non-metals. Like the metals in column 1, they are also very reactive. Metals, for example, react readily with the column 7 non-metals. Magnesium not only burns in air, it will also burn in the gases fluorine and chlorine. When they react with metals, the column 7 non-metals produce salts. Because of this they are called the **halogens**, which means 'salt producer'.

| He |
| helium |
| Ne |
| neon |
| Ar |
| argon |
| Kr |
| krypton |
| Xe |
| xenon |
| Rn |
| radon |

Column 0 – unreactive gases

All of the elements in column 0 are unreactive gases. They are so unreactive that they are called the **noble gases**. Their lack of activity makes them very useful. For example, helium is a safe gas to use in airships (see figure 1.7), argon is used when aluminium is being welded to stop it catching fire, and krypton is used in electric light bulbs (see figure 1.8).

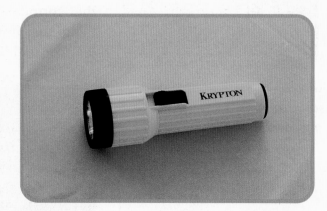

Figure 1.7 *Helium is a safe gas to use in airships*

Figure 1.8 *Krypton is preferred to argon in the bulb of this torch – can you suggest why?*

Section Questions

6 What happens to the shiny surface of a column 1 metal when it is exposed to air?

7 Name the gas that is given off when column 1 metals react with water.

8 Which of the column 7 non-metals is a liquid at room temperature?

9 What property of the column 0 gases makes them suitable for filling electric light bulbs?

Compounds

The previous section dealt only with elements, of which there are only about one hundred. As you look around, you will see thousands of different substances. They cannot all be elements, so what are they? In fact, many of these substances are **compounds**. These are formed when elements react together. Have you ever burned magnesium ribbon? Many people have. When the magnesium burns, it is reacting with oxygen (see figure 1.9). The product of the reaction is a white powder called magnesium oxide.

Metal powders are used in fireworks (see figure 1.10). They give out various colours when they burn, joining with oxygen to make metal oxide compounds.

Many more compounds can be made by elements reacting together. In fact there are millions of compounds all made from about one hundred elements. We can think of elements as being like Lego pieces. With just a few basic types of Lego brick, you can build hundreds of different models (see figure 1.11).

Figure 1.9 *Magnesium and oxygen reacting to produce magnesium oxide*

Figure 1.10 *During firework displays, metals and oxygen react to make metal oxide compounds*

Figure 1.11 *A few basic Lego bricks can be used to build many different models*

Naming compounds

Names often give you information. The name 'Norman' means 'man from the north'. The name 'McDonald' means 'son of Donald'.

Most compounds with the name ending -**ide** contain *two* elements. For example:

- magnesium and oxygen react to give the compound magnesium ox**ide**,
- sodium and chlorine react to give the compound sodium chlor**ide**,
- copper and sulphur react to give the compound copper sulph**ide**.

Notice that when a metal reacts with a non-metal, it is the ending of the non-metal that is changed to -ide. Some more examples of compounds containing two elements are given in table 1.1.

name of compound	elements present
potassium chloride	potassium and chlorine
aluminium bromide	aluminium and bromine
silver iodide	silver and iodine
copper oxide	copper and oxygen
mercury sulphide	mercury and sulphur

Table 1.1 *Some compounds containing two elements*

Many compounds contain more than two elements, with oxygen being present in a lot of them. Chemists have a special way of showing the presence of oxygen. They do this by the name endings -**ate** and -**ite**. In magnesium sulphate, for example, the elements present are magnesium, sulphur and oxygen. In magnesium sulphite, the same elements are present, but there is less oxygen. Some more examples are given in table 1.2.

name of compound	elements present
calcium carbonate	calcium, carbon and oxygen
sodium nitrite	sodium, nitrogen and oxygen
lead phosphate	lead, phosphorus and oxygen

Table 1.2 *Some compounds containing three elements*

Section Questions

10 Give the names of the elements present in the following compounds:
   a) hydrogen oxide,
   b) zinc iodide,
   c) tin sulphide.

11 Give the names of the elements present in the following compounds:
   a) magnesium nitrate,
   b) lithium sulphite,
   c) copper carbonate.

Mixtures

A **mixture** is formed when two or more substances come together but do not react. Many natural substances are mixtures. Air and sea water are examples of mixtures.

Air is a mixture of many gases including nitrogen, oxygen and carbon dioxide.

> The main gases present in air are nitrogen (about 80%) and oxygen (about 20%).

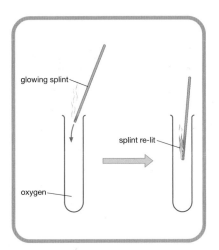

Figure 1.12 *Oxygen will relight a glowing splint*

Oxygen is the gas that lets things burn. When a glowing splint is placed in a test tube of oxygen, it relights (see figure 1.12). Oxygen is the only common gas that can do this. This is therefore used as the chemical test for oxygen.

Although a glowing splint may continue to glow in air, there is not enough oxygen for it to relight.

Mixtures are usually quite easy to separate. For example, the salts in sea water can be obtained by simply evaporating off the water. In hot countries, the sun's rays can be used for this (see figure 1.13).

Figure 1.13 *The sun's rays can be used to evaporate off the water from sea water*

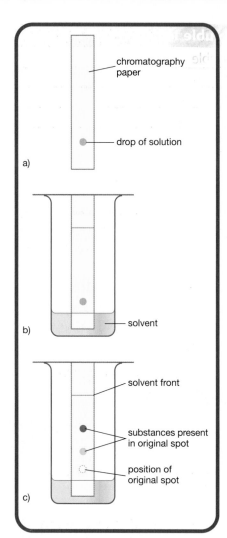

Figure 1.14
Chromatography can be used to separate the coloured substances in food colours and inks

Food colours and the inks in coloured pens are mixtures, from which the coloured substances can be easily separated. The technique used is called **chromatography**. A drop of mixture is spotted onto absorbent paper as shown in figure 1.14 and the bottom of the paper is placed in water.

Many common substances are mixtures. Here is a list of some of them. Can you think of any others?

milk
cup of tea
cup of coffee
soft drinks
alcoholic drinks
natural gas
petrol
diesel
shampoos
washing-up liquid

Solutions

When sugar dissolves in tea or coffee, it 'disappears' (see figure 1.15). We cannot see it, but it must still be there because we can taste it. A special kind of mixture has been formed called a **solution**.

Figure 1.15 *A sugar solution about to be made*

A solution is formed whenever a substance dissolves in a liquid. The most common liquid is of course water. When a substance like sugar dissolves in a liquid such as water, we say that the substance is **soluble**. Substances, like sand, that do not dissolve are said to be **insoluble**. Table 1.3 shows the results of testing the solubility of some common substances in water.

substance	soluble in water?
sugar	soluble
salt	soluble
sand	insoluble
chalk	insoluble

Table 1.3 *Solubilities of some common substances*

Section Questions

12 Say whether the following are mixtures or compounds:
   a) air,
   b) pure water,
   c) milk,
   d) cup of tea.

13 How would you test a gas to see if it was oxygen?

14 Use the table of solubilities on page 4 of the data booklet to give the solubilities of the following compounds:
   a) calcium nitrate,
   b) iron carbonate,
   c) potassium sulphate.

Solutions can be described as **dilute** or **concentrated** depending on how much substance is dissolved in them. A **dilute solution** contains only a little substance compared to the amount of liquid. A **concentrated solution** contains a lot of substance compared to the amount of liquid. Not many people like drinking concentrated orange juice. We make it more dilute (see figure 1.16) before we drink it by adding water.

Figure 1.16 *A concentrated solution and a dilute solution*

CuSO

If you add a little copper sulphate to water, the dilute solution produced is pale blue (see figure 1.17). Adding more copper sulphate makes the solution more concentrated and the colour darkens. Eventually, no more copper sulphate dissolves. When this happens, we say that a **saturated solution** has been formed. This is a solution in which no more substance can be dissolved.

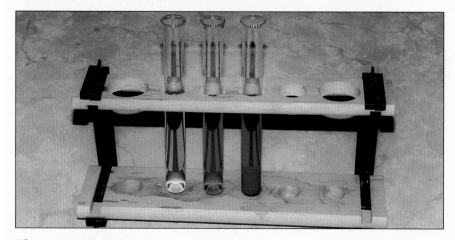

Figure 1.17 *Dilute, concentrated and saturated copper sulphate solutions*

Making drinks fizzy

Drinks can be made fizzy by dissolving the gas carbon dioxide in them. Most bottles and cans of soft drinks have labels on them showing that they contain 'carbonated water'. This means that they contain carbon dioxide (see figure 1.18).

Figure 1.18 *Irn Bru contains carbon dioxide*

We can carry out a special test to show that fizzy drinks contain carbon dioxide. Remove about half of the contents of a bottle of a fizzy drink and then put the top back on and leave it for a while. Will the remaining drink be as fizzy? No – it goes flat! This is because gas has escaped from the liquid and is now at the top of the bottle. If you now remove some of this gas with a syringe and pass it slowly into lime water, you will see the lime water turn milky. This is the special test for carbon dioxide. The diagrams in figure 1.19 show how the experiment is carried out.

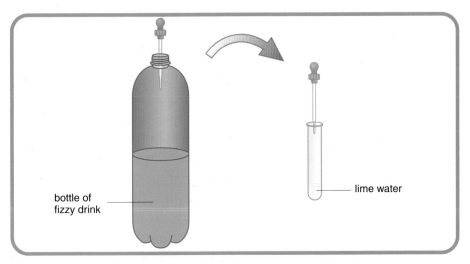

bottle of
fizzy drink

lime water

Figure 1.19 *Showing that a fizzy drink contains carbon dioxide gas*

What's in our water?

Would you drink water out of a river? Perhaps not – it might contain harmful germs and bacteria. However, in many areas, water for use by the public is taken from rivers. The good thing is that, before it reaches our homes, **chlorine** is added to it. Chlorine kills germs and bacteria in the water, making it fit for us to drink.

Surveys have shown that where you live can affect the resistance of your teeth to decay. In areas where **fluoride** compounds are found naturally in the water supply, teeth tend to be more healthy. Fluoride compounds are known to strengthen tooth enamel, the outside layer of the tooth. In some areas, sodium fluoride is added to the water supply. Also, if you want to give your teeth that extra bit of protection, you can buy toothpaste containing fluoride compounds (see figure 1.20).

Why could water in an old house be harmful? One of the reasons could be that the house still has water pipes made of **lead**. Water passing through these can result in lead compounds dissolving in the water. These are poisonous and can be harmful to health. Any lead pipes should be replaced by ones made of copper or plastic.

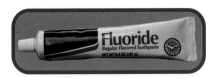

Figure 1.20 *Fluoride compounds help to prevent tooth decay*

The effect of temperature changes on dissolving speed

Information

The aim of this experiment is to find out how changing the temperature of water changes the speed at which sugar dissolves.

What to do

1 Fill a glass beaker about half full of water.
2 Use a syringe to put water from the beaker into a test tube. Do this until the water is about 3 cm from the top.
3 Fill a spatula with sugar crystals and put them in the test tube with the water. Put a stopper on the test tube.
4 Turn the test tube upside down. Hold it upside down until the crystals fall to the bottom. This is called one 'upturn'. Then turn the test tube the right way up again until the crystals fall to the bottom. This counts as the second 'upturn' (see figure 1.21).
5 Keep on doing this. Record the number of 'upturns' you have to do before the crystals just disappear and dissolve completely.
6 Take the stopper out of the test tube and measure the temperature of the solution using a thermometer.
7 Put the beaker of water on a tripod and gauze. Light a bunsen burner and put it under the tripod (see figure 1.22). Heat the beaker of water until the temperature reaches between 35 and 40°C. Remove the bunsen burner.
8 Use a syringe to put warm water from the beaker into a test tube. Add this until it is about 3 cm from the top.
9 Measure the temperature of the water.
10 Fill a spatula with the same amount of sugar crystals as before. Put them in the test tube and put a stopper on the test tube.
11 Again count the number of 'upturns' you have to do until the sugar crystals just dissolve and disappear. Work at the same speed as before.
12 Measure the temperature of the solution in the test tube.
13 Repeat the experiment again after heating the water to between 55 and 65°C.
14 Some typical results are shown in the table.

experiment	1	2	3
first temperature / °C	19	31	47
second temperature / °C	21	29	43
average temperature / °C	20	30	45
number of 'upturns'	16	12	6

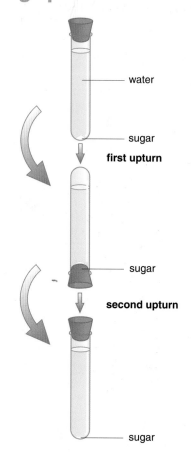

Figure 1.21 *Two 'upturns'*

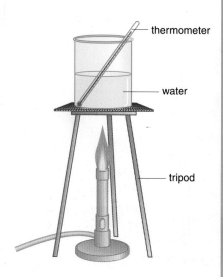

Figure 1.22 *Heating the water*

As the temperature increases, the speed at which the sugar dissolves also increases.

In any investigation only one variable is allowed to change. In this case it is the temperature of the water.

Making the experiment better

The amount of sugar could be kept the same more accurately by weighing out equal masses, such as 1 g, each time. Can you think of any other ways of making the investigation fairer?

Figure 1.23 *Bleach is just one of many hazardous substances in a house*

Figure 1.24 *This lorry is carrying a toxic substance*

Hazard symbols

Can you think of any everyday substance used in the home that is harmful? You might think of something like bleach. Bleach contains chemicals that are used to kill germs and bacteria, so it is probably harmful to us as well. Look at figure 1.23 and you will see that bottles of bleach have a hazard warning symbol on them.

There are four common hazard symbols:

 This symbol shows that a chemical is **harmful** or **irritant**. Bleach and dishwasher powder come into this category.

 The skull and cross-bones are a clue here. Chemicals with this symbol can kill us. They are **toxic**, meaning that they are **poisonous**. Lead compounds are poisonous.

 Can you guess from the flame what this means? You would find this symbol on a can of petrol or paraffin. It means that the substance can catch fire. In other words, it is **flammable**.

 As the symbol suggests, these chemicals cause severe burns on the skin. They are said to be corrosive and can also make holes in metals. A lot of acids are **corrosive**.

By law lorries that carry harmful substances must have special labels showing what the hazard is (see figure 1.24). This means that, in the event of an accident, the emergency services will know how to deal with any spillage.

In school laboratories, hazard labels are attached to all appropriate chemicals.

ACCESS 3 Subsection Test: Substances

Part A

This part of the paper consists of four questions and is worth 4 marks.

1 Which of the following elements is a liquid at room temperature? (1)

   silver or **mercury**

2 Which of the following substances is a mixture? (1)

   air or **water**

3 Which of the following is dissolved in water to kill bacteria? (1)

   chlorine or **sodium fluoride**

4 The hazard symbol for a harmful or irritant substance is: (1)

   OR

   toxic   h

Part B

This part of the paper is worth 6 marks.

5 Dates of discovery of elements are given on page 2 of the data booklet. Give the date of discovery of sodium. (1)

6 When sodium nitrate is heated strongly, a gas is given off, which relights a glowing splint.

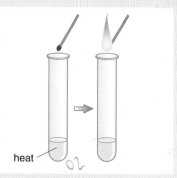

   heat   O2

Name the gas given off in the reaction.   CO2   (1)

7 The bar graph gives information about the use of hydrochloric acid.

50% is used in chemical industries
30% is used in metal industries
15% has other uses
5% is used in food processing

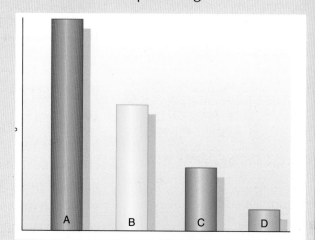

Identify what bar **B** represents. (1)

8 Name the gas that is added to some drinks to make them fizzy.   CO2   (1)

9 The uses of some metals are given on page 5 of the data booklet. Give a use for the metal copper. (1)

10 Information about the solubility of some carbonates and chlorides is given in the table.

	carbonate	chloride
calcium	insoluble	soluble
copper	insoluble	soluble
potassium	soluble	soluble
sodium	soluble	soluble

Use the information in the table to help you to rewrite the following as a correct statement.

Some/all carbonates are insoluble in water, and some/all chlorides are soluble. (1)

Total 10 marks

2 Chemical reactions

Figure 2.1 *Adding vinegar to bicarbonate of soda produces a gas*

Figure 2.2 *Boiling water also produces a gas*

Identification

If you add vinegar to bicarbonate of soda, the mixture fizzes – a gas is given off (see figure 2.1).

If you boil water in a kettle, steam is produced (see figure 2.2).

One of the above is an example of a **chemical reaction**, but which one is it? The answer lies in whether a *new substance* has been formed or not.

When water boils, liquid water changes into steam, but it is still water despite being a gas. It is also easy for the steam to turn back into liquid water. A **physical change** has taken place when water boils.

When vinegar is added to bicarbonate of soda, the gas given off is a new substance called carbon dioxide. Also, it is not possible to turn the new solution and carbon dioxide back into vinegar and bicarbonate of soda. In this case a **chemical reaction** has taken place.

Other evidence of a chemical reaction

When the copper was originally put on the roof of the clock tower shown in figure 2.3, it was a shiny pinkish brown colour. More than sixty years later it has changed colour to green. A *change in appearance*, such as a *colour change*, is evidence of a chemical reaction.

Figure 2.3 *A chemical reaction has caused the copper roof to change colour*

Over the years, copper has reacted with gases in the air to make a new substance. This is copper carbonate, which is green.

A *gas being given off* is also evidence of a chemical reaction (see figure 2.4). For example, adding magnesium to acid produces hydrogen gas in a very rapid reaction. Hydrogen is also a *new substance* produced by the reaction.

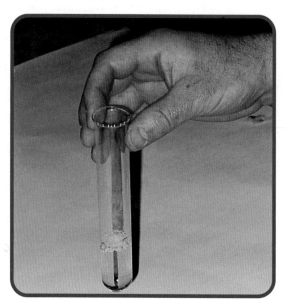

Figure 2.4 *A gas given off is evidence of a chemical reaction*

Sometimes a *solid can form in a clear solution*. This can be a sudden and surprising reaction. For example, colourless solutions of potassium iodide and lead nitrate form the bright yellow solid lead iodide (see figure 2.5). The yellow solid is called a **precipitate**.

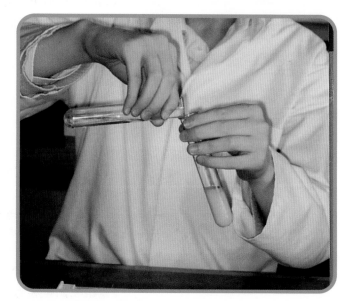

Figure 2.5 *The formation of a precipitate is evidence of a chemical reaction*

Another clue that a chemical reaction may be taking place is an *energy change.* When anything burns (see figure 2.6), like wood or coal, heat energy is given out. This is evidence that a chemical reaction is taking place when something burns. Further evidence is that *new substances* are produced in the form of ash and smoke.

Figure 2.6 *Energy changes are evidence that chemical reactions are taking place*

Some everyday chemical reactions

You might be familiar with some of these chemical reactions:

fuels burning
iron rusting
copper corroding
wine fermenting
bread toasting
eggs frying
cakes baking

Summary

Evidence for a chemical reaction:
- new substance(s) formed
- change in appearance
- colour change
- gas given off
- precipitate formed
- energy change

Section Questions

1 Which of the following are chemical reactions?
   a candle burning
   nail varnish drying
   ice melting
   bacon frying
   water evaporating
   hair bleaching

Speed of chemical reactions

a) Particle size

A log of wood on a fire burns quite slowly. What could have been done to the log to make it burn faster? The answer is to cut it up into smaller pieces. Chemical reactions take place more quickly if *smaller particles* are used.

In the laboratory we can speed up the reaction between a marble chip and acid in a similar way. Take two marble chips of the same size, and then break one up into smaller pieces. Marble and acid react to give carbon dioxide gas. The smaller pieces of marble react faster than the single marble chip (see figure 2.7).

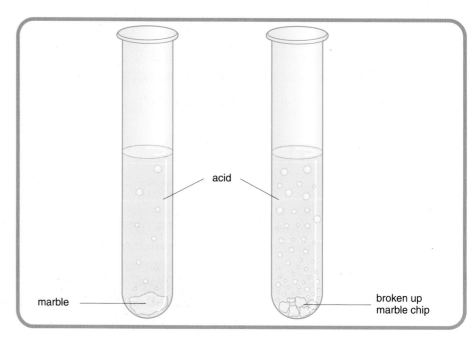

Figure 2.7 *Small pieces of marble react faster with acid than larger pieces*

Figure 2.8 *Why must we store milk at a low temperature?*

b) Temperature

What would happen to some fresh fish if it were left on a work surface in a warm kitchen? Very soon it would start to smell and eventually go rotten. Chemical reactions take place when food goes bad. How would you slow down these reactions in the fish? Where should the fish have been put? The answer is to put the fish (and other foods) in a refrigerator where the low temperature would slow the reactions down (see figure 2.8).

Temperature affects the speed of a chemical reaction. Increasing the temperature also increases the reaction speed. Decreasing the temperature slows the reaction down. Can you tell which of the reaction mixtures in figure 2.9 is at the higher temperature?

Figure 2.9 *Which reaction mixture was at the higher temperature, the test tube on the left, or the one on the right?*

Summary

slow reactions	fast reactions
large particles	small particles
low temperature	high temperature

Section Questions

2 A reaction between a large piece of metal and an acid was found to be very slow. Give *two* ways of increasing the reaction speed.

Catalysts

Sometimes a chemical reaction can be speeded up just by the presence of a special substance called a **catalyst**. A catalyst speeds up a reaction, but is not used up and is still there at the end of the reaction.

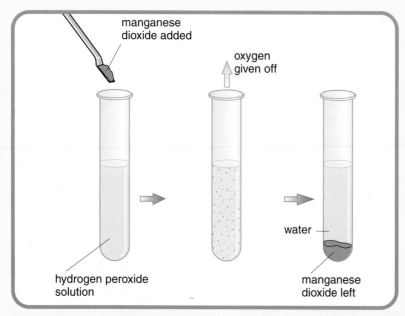

Figure 2.10 *Using a catalyst to speed up a chemical reaction*

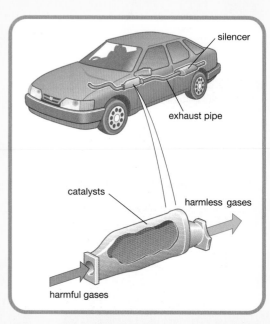

Figure 2.11 *Catalysts are present in the 'catalytic converters' in car exhaust systems*

Figure 2.12 *Platinum acts as a catalyst in this styling brush*

Hydrogen peroxide solution, which is used to bleach hair, slowly breaks down into water and oxygen. The presence of manganese dioxide speeds up this reaction a lot, and it is still there at the end (see figure 2.10). Manganese dioxide is a catalyst for this reaction.

Many catalysts have been discovered. However, *all* catalysts:
- speed up some reactions,
- are not used up during a reaction.

There are catalysts present in some of the everyday things that we use. Some cars have catalysts in their exhaust systems to help break down harmful exhaust gases into harmless ones (see figure 2.11).

One of the catalysts in a 'catalytic converter' is the valuable metal platinum. The same metal acts as a catalyst in some hair styling brushes (see figure 2.12). It catalyses the reaction in which the fuel (inside the handle) burns. The burning fuel heats up the brush.

Enzymes

Enzymes are catalysts that affect living things. They catalyse the chemical reactions that take place in plants and animals. One of these enzymes, called catalase, helps to break down poisonous hydrogen peroxide, found in our bodies, into water and oxygen. Liver contains catalase and, when placed in hydrogen peroxide, causes it to decompose rapidly (see figure 2.13).

Figure 2.13 *Liver contains the enzyme catalase, which breaks down hydrogen peroxide*

Figure 2.14 *Soft centres are made soft by an enzyme*

Vegetables like potatoes also contain catalase. However, when pieces of these are added to hydrogen peroxide, the decomposition is slower.

Have you ever thought how difficult it must be to coat the soft centre of a chocolate with molten chocolate? Why does the chocolate not collapse? The answer is that a harder centre is made using sucrose as the sugar present. Molten chocolate is poured over this to give a coating. The centre then softens because the enzyme invertase was mixed with the sucrose. This catalyses the breakdown of the sucrose into two simpler sugars, which makes the centre softer (see figure 2.14).

Enzymes are added to washing powders because they can break down stains like blood and fat (see figure 2.15).

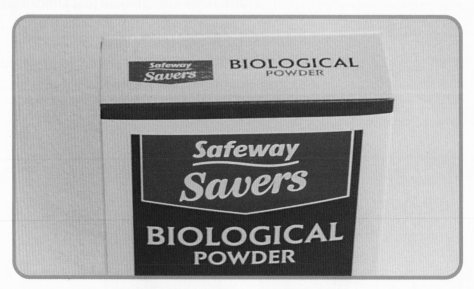

Figure 2.15 *Some washing powders contain enzymes*

Yeast provides a variety of enzymes, which are used in making whisky, beer and lager (see figure 2.16).

Figure 2.16 *These drinks are made using enzymes*

Section Questions

3 A gas is given off when the enzyme catalase in liver breaks down hydrogen peroxide. How would you show that this gas is oxygen. Describe the test and give the result.

The effect of concentration changes on reaction speed

Information

The aim of this experiment is to find out how changing the concentration of sulphuric acid changes the speed with which it reacts with magnesium.

What to do

1 Using a syringe, add 20 cm³ of sulphuric acid with a concentration of 2 mol/l to a small beaker (see figure 2.17).
2 Add a 2 cm piece of magnesium ribbon to the acid and start a timer.
3 When the magnesium stops fizzing and has disappeared, stop the timer and note the time on it.
4 Wash and dry the small beaker.
5 Add 10 cm³ of the 2 mol/l sulphuric acid to the beaker along with 10 cm³ of water. This gives an acid concentration of 1 mol/l.
6 Add a 2 cm strip of magnesium ribbon to the acid and note the time taken for the magnesium to react and disappear.
7 Repeat the experiment using 5 cm³ of 2 mol/l acid and 15 cm³ of water. This gives an acid concentration of 0.5 mol/l.

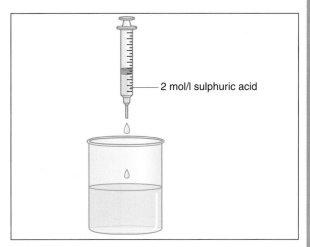

Figure 2.17 *Adding acid to the beaker*

2 mol/l sulphuric acid

Some typical results are shown in the table.

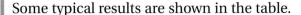

experiment	1	2	3
acid concentration / mol/l	2	1.0	0.5
reaction time / s	12	22	45

As the acid concentration increases, the reaction speed increases.

What would happen if ...?

Sometimes bubbles stick to the magnesium ribbon causing it to float. What effect would this have on reaction time? Would it make the reaction faster or slower?

Word equations

So far we have used sentences to describe chemical reactions. For example, calcium and oxygen react to produce calcium oxide. Chemists have a short-hand way of giving this information in what is called a **word equation**. For example:

calcium + oxygen → calcium oxide

In all word equations, '+' means 'and'. The arrow '→' means 'react to produce'. So the above word equation tells us that 'calcium and oxygen react to produce calcium oxide'.

Calcium and oxygen are said to be the **reactants**, as they are the chemicals involved in the reaction. Calcium oxide is said to be the **product** of the reaction, as it was made by the reaction.

When you use a bunsen burner to heat something, you are using the heat given out by a chemical reaction. If North Sea gas is the fuel in your area, then the flame of the bunsen burner is where methane and oxygen react to produce carbon dioxide and water (see figure 2.18).

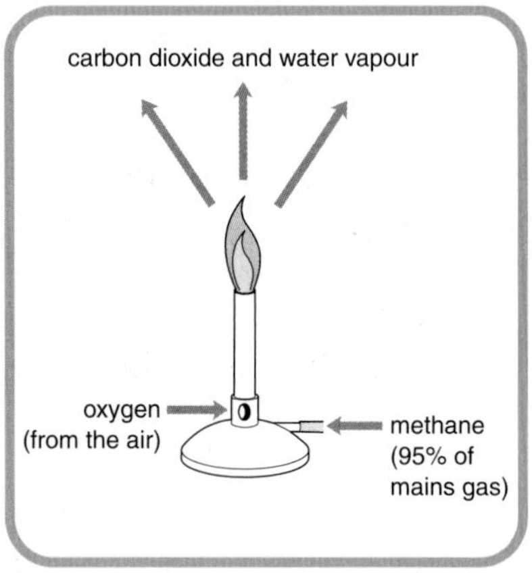

The word equation for the reaction taking place in the bunsen flame is

methane + oxygen → carbon dioxide + water

carbon dioxide and water vapour

oxygen (from the air)

methane (95% of mains gas)

Figure 2.18 *In a bunsen flame, methane and oxygen react to produce carbon dioxide and water*

Section Questions

4 Write word equations for the following chemical reactions.
  a) Sodium reacts with chlorine to produce sodium chloride.
  b) Potassium and water react to produce potassium hydroxide and hydrogen.
  c) Copper sulphate and water are produced when copper oxide and sulphuric acid react. (Read this question again carefully!)
  d) When heated, silver oxide reacts to produce silver and oxygen.

ACCESS 3 Subsection Test: Chemical reactions

Part A

This part of the paper consists of four questions and is worth 4 marks.

1 Which of the following is an example of a chemical reaction? (1)

   ice melting or **a match burning**

2 Which of the following would react faster with dilute acid? (1)

   zinc powder or **zinc lumps**

3 Chalk reacts with hydrochloric acid. In which case would the reaction be slower? (1)

   acid at 30°C or **acid at 10°C**

4 What is the *smallest* number of new substances that can be formed in a chemical reaction? (1)

   one or **two**

Part B

This part of the paper is worth 6 marks.

5 Marble chips and dilute acid react to produce carbon dioxide gas. In the experiment shown below, the mass of carbon dioxide given off was noted at equal time intervals.

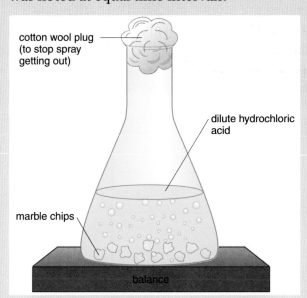

cotton wool plug (to stop spray getting out)

dilute hydrochloric acid

marble chips

balance

time / seconds	mass of gas / grams
0	0
30	2.0
60	3.0
90	3.6
120	3.9
150	4.0

a) What was the time interval between each measurement? (1)

b) What mass of gas had been given off after 90 seconds? (1)

6 When green copper carbonate is heated it turns black and a gas is given off.

a) Give **two** pieces of evidence to show that a chemical reaction has taken place? (1)

When heated, copper carbonate reacts to produce copper oxide and carbon dioxide.

b) Write a word equation for this chemical reaction. (1)

7 When solutions of sodium chloride and silver nitrate are mixed together, a white precipitate forms very quickly.

Are the particles that react likely to be very large or very small? (1)

8 Which of the following is likely to give the fastest chemical reaction? (1)

	particle size	temperature
A	large	high
B	small	low
C	large	low
D	small	high

Total 10 marks

3 Bonding

Atoms

Every element is made up of very small particles called **atoms**. A gold ring is made up of gold atoms. A copper pipe is made up of copper atoms. The atoms of gold and copper are different. For example, gold atoms are heavier than copper atoms. In fact, each of the hundred or so elements is made up of different atoms.

Each element is given a special number, called the **atomic number**, which puts the elements in order. As the atomic number increases, the mass of the atoms also tends to increase. The atomic numbers of the first ten elements in the Periodic Table are as shown in table 3.1.

element	atomic number
hydrogen	1
helium	2
lithium	3
beryllium	4
boron	5
carbon	6
nitrogen	7
oxygen	8
fluorine	9
neon	10

Table 3.1 *Atomic numbers for the first ten elements in the Periodic Table*

You will probably use boxes of atomic models. The hard plastic spheres have different colours and sizes. This is because they represent atoms of different elements. Some are shown in table 3.2.

element	atomic model
hydrogen	
carbon	
nitrogen	
oxygen	
chlorine	

Table 3.2 *Coloured plastic spheres are used to represent atoms*

Molecules

Atoms of the elements shown in the tables and others can join together in clusters called **molecules**. The clusters are made up of two or more atoms held together by strong **bonds**. Molecules of the element hydrogen contain just two atoms (see figure 3.1).

Chemical formulae tell us how many atoms are present in a molecule of a substance. The chemical formula for hydrogen is H_2.

Molecules of nitrogen, oxygen and chlorine also contain just two atoms of each element (see table 3.3).

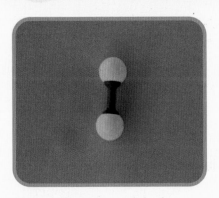

Figure 3.1 *A model of a hydrogen molecule*

element	molecular model	chemical formula
nitrogen		N_2
oxygen		O_2
chlorine		Cl_2

Table 3.3 *More elements with two atoms in each of their molecules*

Many compounds exist as molecules. One of the simplest is hydrogen oxide, the more common name for which is water. Its chemical formula is H_2O. Water and some other compounds that are made up of simple molecules are shown in table 3.4.

compound	molecular model	chemical formula
water		H_2O
ammonia		NH_3
methane		CH_4

Table 3.4 *Some compounds that exist as molecules*

When drawing diagrams of molecules, sometimes chemical symbols are used to represent atoms. A water molecule, for example, can be shown as follows:

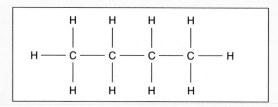

A diagram of a molecule of the fuel butane looks like this:

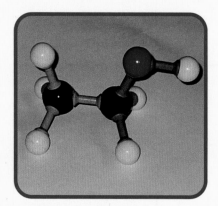

The chemical formula for butane must therefore show that it has four carbon atoms and ten hydrogen atoms in each of its molecules. We write the chemical formula as C_4H_{10}.

The alcohol that is present in beer, wine and spirits like whisky is a small molecule with an interesting shape (see figure 3.2).

A diagram of an alcohol molecule looks like this:

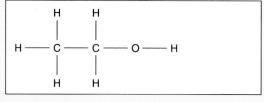

The chemical formula for alcohol can be written as C_2H_6O.

Figure 3.2 *What do you think an alcohol molecule looks like?*

Section Questions

1 Write chemical formulae for the following substances based on the diagrams of their molecules:

a) methanol

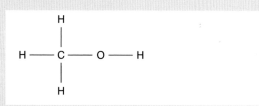

b) propane

c) ethanoic acid

Chemical formulae – from prefixes!

For some compounds, the chemical formula can be worked out from its name. In carbon dioxide, for example, the prefix 'di-' means 'two'. This is because the molecule of carbon dioxide contains one carbon atom and *two* oxygen atoms. The chemical formula for carbon dioxide is therefore CO_2.

There are four **prefixes** that you must know (see table 3.5).

prefix	meaning
mono-	1
di-	2
tri-	3
tetra-	4

Table 3.5 *Prefixes and their meanings*

If no prefix is given, then you must assume that only *one* atom of that element is present in the chemical formula.

Example 1

Carbon monoxide (*mono* = 1)

The chemical formula is CO.

Example 2

Sulphur dioxide (*di* = 2)

The chemical formula is SO_2.

Example 3

Nitrogen triiodide (*tri* = 3)

The chemical formula is NI_3.

Example 4

Silicon tetrachloride (*tetra* = 4)

The chemical formula s $SiCl_4$.

Section Questions

2 Write chemical formulae for the following compounds based on your knowledge of prefixes:
a) nitrogen monoxide,
b) phosphorus tribromide,
c) carbon tetrafluoride,
d) lead dioxide.

Bonds between molecules

The bonds that hold the atoms together inside a molecule are strong. However, the bonds *between* molecules are weak. It is therefore easy to break the bonds between molecules, which is what happens when liquids boil. For example, when water boils (see figure 3.3), molecules that were held close together in liquid water move far apart in steam.

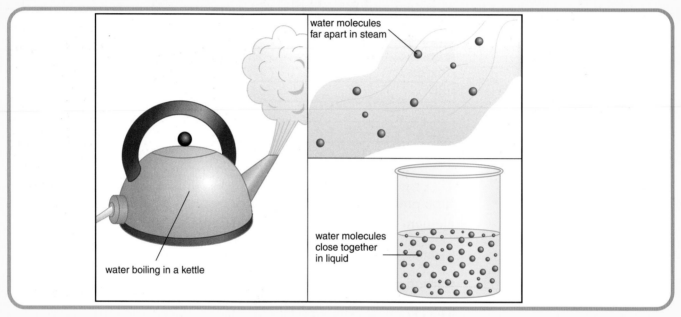

water molecules far apart in steam

water molecules close together in liquid

water boiling in a kettle

Figure 3.3 *When water boils, the weak bonds between molecules are broken*

Some molecular substances are solids at room temperature, but because of the weak bonds between molecules they melt at quite low temperatures. Ordinary sugar, for example, melts at 160°C. The bonds between iodine molecules are even easier to break:

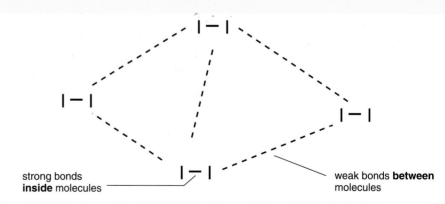

strong bonds **inside** molecules

weak bonds **between** molecules

If you heat some iodine crystals gently, purple iodine vapour forms easily (see figure 3.4).

Figure 3.4 *The heat from hot water is sufficient to break the weak bonds between iodine molecules*

Some melting points and boiling points

If a substance is a gas, a liquid, or a solid with a low melting point, then we may assume that it is made of molecules (see table 3.6). This is because the bonds *between* molecules are weak and easily broken.

name	mp / °C	bp / °C	state
some molecular elements			
chlorine	−101	−35	gas
bromine	−7	59	liquid
iodine	114	184	solid
some molecular compounds			
sulphur dioxide	−73	−10	gas
water	0	100	liquid
naphthalene	81	218	solid

Table 3.6 *Melting points and boiling points for some molecular substances*

Solubility

Most covalent substances do not dissolve much in water. Sugars are unusual in this respect. They do dissolve well in other liquids. Iodine, for example, is almost insoluble in water, but dissolves in alcohol and petrol (see figure 3.5).

Figure 3.5 *Iodine does not dissolve well in (a) water, but does dissolve well in (b) alcohol and (c) petrol*

name	state
propane	gas
butane	gas
water	liquid
nail varnish remover	liquid
petrol	liquid
candle wax	solid
sugar	solid

Table 3.7 *Some common molecular substances*

Conduction of electricity

As a liquid, or in solution, a substance *will* conduct electricity if it is made of electrically charged particles. Covalent substances do not conduct electricity as liquids or in solution because molecules are not electrically charged.

Everyday covalent substances

We have already mentioned several common covalent substances. These have included methane in the gas supplied to homes and carbon dioxide in fizzy drinks. Table 3.7 and figure 3.6 are here to remind you of ones we have met already and to add some new ones. Some, like nail varnish remover, are mixtures of covalent substances. Note that covalent substances can be gases, liquids or solids.

Figure 3.6 *Examples of everyday covalent substances*

Summary

Substances that are made up of molecules:
- have *strong* bonds *inside* the molecules,
- have *weak* bonds *between* the molecules,
- are either gases, liquids, or solids with low melting points,
- do not dissolve well in water,
- do not conduct electricity.

Section Questions

3 Refer to page 4 of the data booklet and give the melting points of the following two molecular solids:
a) naphthalene,
b) phenol.

4 The diagram below shows two water molecules, with two bonds labelled a) and b).

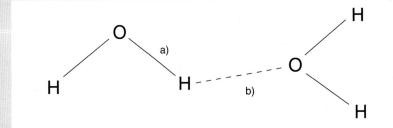

Choose the correct response for a) and b):
a) represents a strong/weak bond.
b) represents a strong/weak bond.

Ions

Some substances are made of electrically charged particles called **ions**. These substances are all compounds that have high melting points (see table 3.8). They are therefore solids at room temperature and the forces *between* the ions must be strong.

ionic compound	mp / °C	state
potassium iodide	686	solid
sodium chloride	801	solid
calcium fluoride	1151	solid

Table 3.8 *Melting points for some ionic compounds*

In the compounds listed in table 3.8, the ions are simply atoms with an electric charge. A small part of sodium chloride is shown below in figure 3.7.

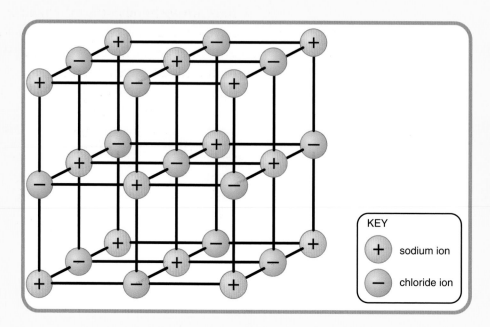

KEY

+ sodium ion

− chloride ion

Figure 3.7 *The way that the ions are arranged in sodium chloride*

Solubility

Many ionic compounds dissolve in water. This is unlike molecular substances, which are mostly insoluble in water. Sometimes the solutions are coloured, as in figure 3.8.

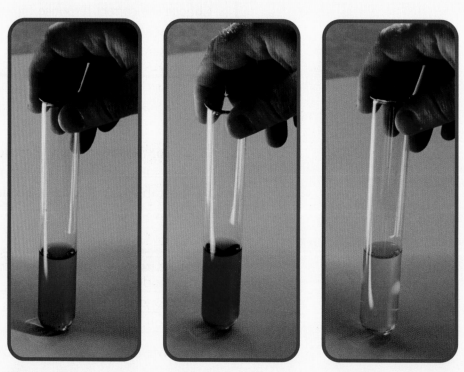

Figure 3.8 *Ionic compounds dissolved to give these coloured solutions*

Conduction of electricity

Solid ionic compounds cannot conduct electricity. Why not? This is because the electrically charged ions in the solid cannot move. When they can move, they can carry an electric current. The ions become free to move when the solid is melted or dissolved in water. It is a lot easier just to dissolve an ionic solid in water than to melt it. (This is because the melting points are high.) We can therefore use the apparatus in figure 3.9 to find out if a compound is made of ions or not. If it is made of ions, the solution will conduct electricity and the bulb will light up.

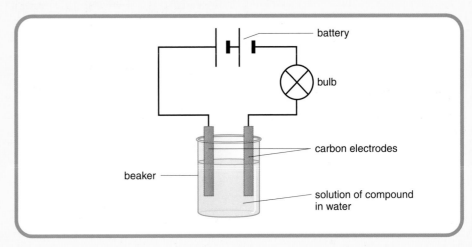

Figure 3.9 *A solution of an ionic compound conducts electricity*

Everyday ionic compounds

Sodium chloride has a common, everyday name. It is called 'common salt', or simply 'salt'. This and some other common ionic compounds are shown in table 3.9 and figure 3.10. Note that they are all solids.

Figure 3.10 *Examples of everyday ionic compounds*

name	state
salt	solid
bicarbonate of soda	solid
washing soda	solid
lime (for the garden)	solid
weed killer	solid
indigestion tablets	solid

Table 3.9 *Some common ionic compounds*

Summary

Substances that are made up of ions:
- have *strong* bonds between the ions,
- are all solids,
- are all compounds,
- tend to dissolve well in water,
- conduct electricity when dissolved in water or when molten.

Section Questions

5 Refer to page 4 of the data booklet and give the melting points of the following ionic compounds:
a) potassium fluoride,
b) calcium oxide.

6 Copy and complete the following by choosing the correct words:
All ionic compounds are solid/liquid at room temperature because they have weak/strong bonds between the ions.

7 How would you use the apparatus in figure 3.9 to show that bicarbonate of soda is an ionic compound?

4 Acids and alkalis

Figure 4.1 *Universal indicator and pH paper give pH values by colour matching (see figure 4.3)*

Figure 4.2 *pH meters give accurate pH values*

The pH scale

You will find **acids** and **alkalis** at home and in the laboratory. You can also find substances that are neither acids nor alkalis. These substances are said to be **neutral**.

How can you tell which is which? This is done by measuring something called the **pH value** of a solution of the substance in question. There are three ways of measuring the pH values of solutions. Universal indicator and pH paper give a colour that can be matched against a standard chart of pH values (see figure 4.1). A more accurate pH value can be obtained using a pH meter (see figure 4.2).

The pH scale ranges from below 0 to above 14. Acids have a pH of less than 7, pure water and neutral solutions have a pH of 7, and alkalis have a pH of more than 7 (see table 4.1).

solution	pH
acid	less than 7
neutral	equal to 7
alkali	more than 7

Table 4.1 *The pH scale*

There are various types of Universal indicator and pH paper, but one of the most common follows the colours of the rainbow. These colours and the pH values that they correspond to are shown in figure 4.3.

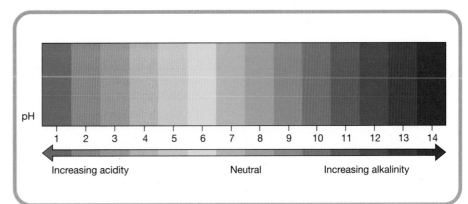

Figure 4.3 *Colours and pH values for a common Universal indicator and pH paper*

41

As the chart in figure 4.3 shows, the lower the pH of an acid, the greater is the acidity. For example, a dilute solution of hydrochloric acid with a pH value of 1 is more acidic than vinegar with a pH value of 3.

In a similar way, we can see that the higher the pH of an alkali, the greater is the alkalinity. A solution of sodium hydroxide with a pH of 14 is therefore more alkaline than a solution of ammonia with a pH of 11.

What should you do if you spill an acid or an alkali on yourself? Probably the best answer is to wash it off with water immediately (see figure 4.4). The reason why this is a good idea is because water is neutral, so it dilutes the acid or alkali. By doing this, it makes an acid less acidic and an alkali less alkaline. In both cases the pH moves towards 7.

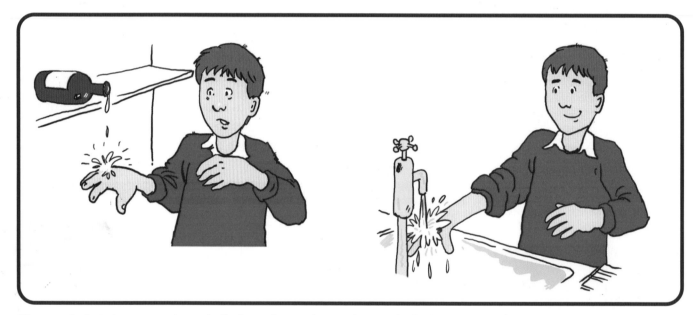

Figure 4.4 *Diluting acids and alkalis reduces the acidity and alkalinity*

Section Questions

1 What is the pH value of a neutral solution?

2 One solution has a pH value of 2 and another has a pH value of 5. Which is the more acidic?

3 One solution has a pH value of 12 and another has a pH value of 9. Which is the more alkaline?

4 Copy and complete the following by choosing the correct words:
   a) When an acid solution is diluted with water, the pH falls/rises and the acidity increases/decreases.
   b) When a solution of an alkali is diluted with water, the pH falls/rises and the alkalinity increases/decreases.

Prescribed Practical Activity

Testing the pH of a solution

Information

The aim of this experiment is to find the pH values of some household substances and to classify them as acidic, alkaline or neutral.

This experiment can be carried out using either pH indicator solution or pH paper.

What to do

Using pH indicator solution

1 Add some vinegar to a test tube to a depth of about 2 cm (see figure 4.5).

2 Add 2 or 3 drops of pH indicator solution to the vinegar and shake the test tube (see figure 4.6).

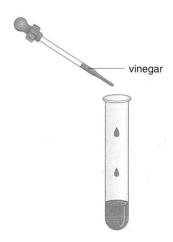

Figure 4.5 *Adding the vinegar*

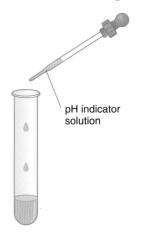

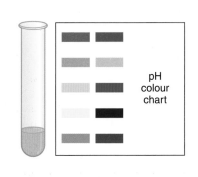

Figure 4.6 *Adding the indicator*

Figure 4.7 *Matching the colour*

3 Match the colour of the solution to one on the corresponding pH colour chart. This will give the pH of the vinegar (see figure 4.7).

4 Repeat the three steps 1–3 above using lemon juice in place of vinegar.

5 Then repeat the three steps 1–3 with soda water, and finally with diluted household ammonia.

6 In order to test a solid like common salt, it must first be made into a solution.

7 Add some water to a test tube to a depth of about 2 cm. Use a spatula to add a small amount of common salt and shake the mixture to make a solution.

8 Add 2 or 3 drops of pH indicator solution and shake the test tube.

9 Match the colour of the solution to one on the corresponding pH colour chart. This will give the pH of the common salt solution.

10 Repeat the three steps 7–9 above with bicarbonate of soda instead of salt.

11 Then repeat the three steps 7–9 with sugar, and finally with automatic washing powder.

What to do

Using pH paper

1 Add a few drops of vinegar to a dimple in a dimple tray (see figure 4.8).

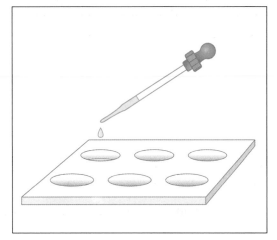

Figure 4.8 *Adding vinegar to a dimple*

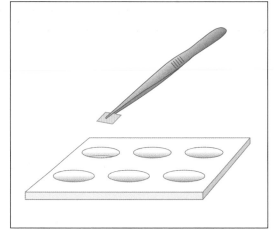

Figure 4.9 *Testing with pH paper*

2 Using tweezers, dip a small piece of pH paper into the vinegar (see figure 4.9).

3 Match the colour of the pH paper to one on the corresponding pH colour chart (see figure 4.10).

4 Repeat the three steps 1–3 above using lemon juice in place of vinegar.

5 Then repeat the three steps 1–3 with soda water, and finally with diluted household ammonia.

6 In order to test a solid like common salt, it must first be made into a solution.

7 Add a few drops of water to a dimple in the dimple tray. Use a spatula to add a very small amount of common salt to the water.

8 Using tweezers, dip a small piece of pH paper into the salt solution.

9 Match the colour of the pH paper to one on the corresponding pH colour chart.

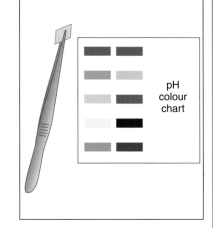

pH colour chart

Figure 4.10 *Matching the colour*

10 Repeat the three steps 7–9 above with bicarbonate of soda instead of salt.

11 Then repeat the three steps 7–9 with sugar, and finally with automatic washing powder.

Some typical results are shown in the table.

substance	pH	acidic/alkaline/neutral
vinegar	4	acidic
lemon juice	3	acidic
soda water	5	acidic
diluted household ammonia	10	alkaline
common salt	7	neutral
bicarbonate of soda	9	alkaline
sugar	7	neutral
automatic washing powder	11	alkaline

Something to think about

Sand is acidic. It reacts with alkalis. However, we cannot measure the pH of sand. Why not?

Common acids and alkalis

Acids and alkalis are in common use in the home, in industry and in the laboratory.

Common laboratory acids include hydrochloric acid, sulphuric acid and nitric acid (see table 4.2). These are all very acidic, with low pH values, and all are labelled as 'corrosive' (see figure 4.11).

name of acid	formula	acidity
hydrochloric acid	HCl	high
sulphuric acid	H_2SO_4	high
nitric acid	HNO_3	high

Table 4.2 *Common laboratory acids*

Figure 4.11 *Which of these substances contains the strongest acid?*

Common laboratory alkalis include sodium hydroxide, lime water and ammonia solution (see table 4.3). Lime water is a dilute solution of calcium hydroxide and, like ammonia solution, is moderately alkaline. All solutions of sodium hydroxide are very alkaline, and have high pH values. Solutions of sodium hydroxide are labelled 'corrosive' (see figure 4.12). Lime water and laboratory solutions of ammonia are usually labelled as 'harmful' (see figure 4.12).

name of alkali	formula	alkalinity
sodium hydroxide	NaOH	high
lime water	$Ca(OH)_2$	moderate
ammonia	NH_3	moderate

Table 4.3 *Common laboratory alkalis*

Figure 4.12 *Which of these substances contains the strongest alkali?*

Common household acids include vinegar, lemonade, soda water and Coca Cola. Not surprisingly, these substances are only moderately acidic. Vinegar, for example, has a pH value of about 3. The acids present in these substances are shown in table 4.4.

substance	acid present
vinegar	ethanoic acid
lemonade	citric acid
soda water	carbonic acid
Coca Cola	phosphoric acid

Table 4.4 *Common household acids*

There are many other household acids, as you can see in figure 4.13.

All acids have a sour taste (see figure 4.14). What do you think is added to drinks like lemonade to improve the taste? The answer, of course, is sugar or artificial sweeteners. So-called 'diet' drinks contain no sugar, only artificial sweeteners.

Figure 4.13 *All of these products contain acids*

Figure 4.14 *Which juice contains most acid?*

Common household alkalis include baking soda, oven cleaner, dishwashing powder, bleach and soaps (see table 4.5). The alkalinity of oven cleaner, dishwashing powder and bleach is sufficiently high that they carry 'irritant' labels (see figure 4.15). You are probably aware that soap can also be 'irritant' if some gets in your eye! Baking soda is only slightly alkaline, and does not merit a hazard label.

substance	alkali present
baking soda	sodium hydrogencarbonate
oven cleaner	sodium hydroxide
dishwashing powder	sodium carbonate
bleach	sodium hydroxide
soap	sodium stearate

Table 4.5 *Common household alkalis*

You will be able to spot some other household alkalis in figure 4.15.

Alkalis have an unpleasant taste. Only ones with low alkalinity are consumed by humans. For example, baking soda, which is also known as 'bicarbonate of soda', is sometimes found in toothpaste and in indigestion remedies.

Figure 4.15 *All of these products contain alkalis*

The pH values for some of the acids and alkalis referred to in this section are shown in figure 4.16.

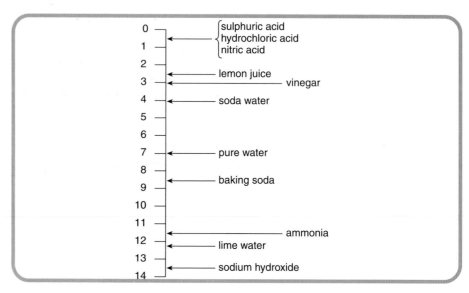

0	sulphuric acid
1	hydrochloric acid
	nitric acid
2	
3	lemon juice, vinegar
4	soda water
5	
6	
7	pure water
8	baking soda
9	
10	
11	ammonia
12	lime water
13	sodium hydroxide
14	

Figure 4.16 *The pH values of some substances*

Section Questions

5 Name two common laboratory acids.

6 Name two common laboratory alkalis.

7 Name three common household acids.

8 Name three common household alkalis.

9 Refer to figure 4.16 and give pH values for:
  a) vinegar,
  b) soda water,
  c) baking soda,
  d) ammonia.

Neutralisation

The Swedish lake shown in figure 4.17 is acidic. Fish cannot survive in water that is too acidic. The helicopter is dropping lime into the lake, because lime makes the water less acidic, so fish do not die. Lime is an alkali and it reacts with acids in a **neutralisation** reaction.

All alkalis react with acids and neutralise them to form a salt and water. We can write a general word equation for this reaction:

alkali + acid → salt + water

Figure 4.17 *Why is lime being dropped into this Swedish lake?*

For example, the alkali sodium hydroxide reacts with hydrochloric acid to form sodium chloride and water. Sodium chloride is better known as common salt and it is neutral. So if sodium hydroxide is added to hydrochloric acid with a pH value of 1, sodium chloride with a pH value of 7 is formed. Whenever an alkali is added to an acid, neutralisation moves the pH *up* towards 7 (see figure 4.18).

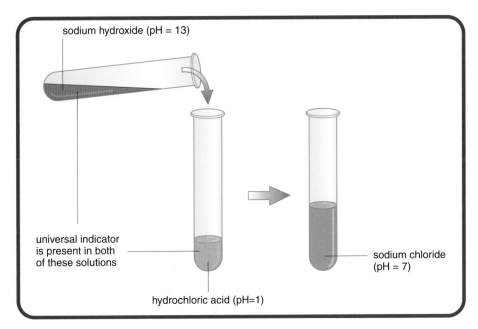

Figure 4.18 *Neutralisation of an acid moves the pH up towards 7*

If an acid is added to an alkali, neutralisation moves the pH *down* towards 7 (see figure 4.19).

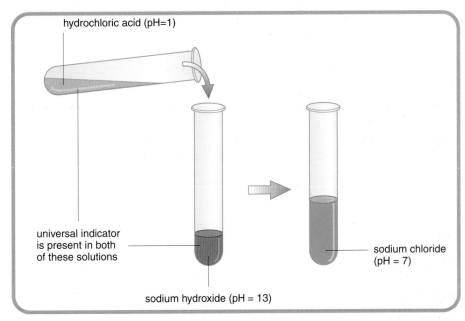

Figure 4.19 *Neutralisation of an alkali moves the pH down towards 7*

Making a salt by neutralisation

We could separate the common salt, sodium chloride, from either of the final solutions shown in figures 4.18 and 4.19. Unfortunately, there would be a problem. The salt would be coloured green by the Universal indicator. How could we make a salt by neutralisation that was not coloured?

One way of obtaining a salt by neutralisation that is not coloured by Universal indicator is shown in figure 4.20.

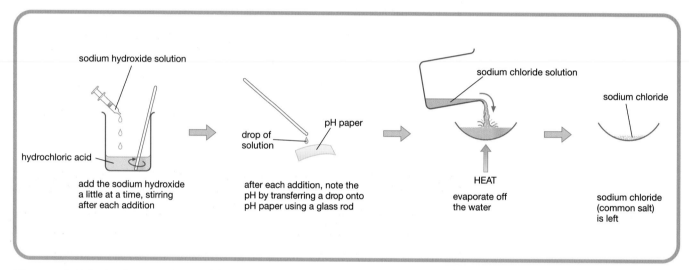

sodium hydroxide solution

sodium chloride solution

sodium chloride

pH paper

drop of solution

hydrochloric acid

add the sodium hydroxide a little at a time, stirring after each addition

after each addition, note the pH by transferring a drop onto pH paper using a glass rod

HEAT

evaporate off the water

sodium chloride (common salt) is left

Figure 4.20 *Making sodium chloride crystals by neutralisation followed by evaporation*

Section Questions

10 Apart from water, what is formed when an acid reacts with an alkali?

11 How can the water be removed from a solution of a substance in water?

12 Copy and complete the following by choosing the correct word:
a) Neutralisation moves the pH of an acid up/down towards 7.
b) Neutralisation of an alkali moves the pH up/down towards 7.

Naming salts

We have seen already that alkalis and acids react to produce a salt and water:

alkali + acid → salt + water

The name of a salt can be worked out from the names of the alkali and acid that react to produce it. There are two parts to the name of a salt. The first part is the metal in the alkali. For example, **sodium** hydroxide produces a **sodium salt**. The second part comes from the acid. For example, **hydrochloric** acid produces a **chloride salt**.

Therefore, **sodium** hydroxide and **hydrochloric** acid react to produce **sodium chloride**. The complete word equation for the reaction is

sodium hydroxide + hydrochloric acid → sodium chloride + water

Tables 4.6 and 4.7 give the names of salts produced by some alkalis and acids.

name of alkali	name of salt
sodium hydroxide	sodium …………
potassium hydroxide	potassium …………
calcium hydroxide	calcium …………

Table 4.6 *Naming salts from alkalis*

name of acid	name of salt
hydrochloric acid	………… chloride
sulphuric acid	………… sulphate
nitric acid	………… nitrate

Table 4.7 *Naming salts from acids*

Example 1

Potassium hydroxide and sulphuric acid react to produce the salt potassium sulphate.

Example 2

Calcium hydroxide and nitric acid react to produce the salt calcium nitrate.

Section Questions

13 Give the name of the salt that is produced by the reaction between each of the following:
a) sodium hydroxide and sulphuric acid,
b) potassium hydroxide and nitric acid,
c) calcium hydroxide and hydrochloric acid.

14 Copy and complete the following word equations:
a) calcium hydroxide + sulphuric acid → .... + water
b) sodium hydroxide + ... → sodium nitrate + water
c) .... + nitric acid → calcium nitrate + water

Neutralising acids with metal carbonates

Farmers use sulphuric acid to kill off the tops of potato plants before harvesting the potatoes. The acid would leave the soil unfit for growing more crops if it were not neutralised first. What would you neutralise the acid soil with? Farmers in Scotland use powdered limestone (see figure 4.21). Limestone is mainly calcium carbonate, which, like all metal carbonates, can neutralise acids.

Figure 4.21 *Crushed limestone is used to neutralise soil acidity*

Figure 4.22 *Carbonate rocks react with acids to produce carbon dioxide gas*

When metal carbonates neutralise acids, there is a fizz – the gas carbon dioxide is given off. Can you see the fizz produced when acid was added to a piece of limestone in figure 4.22?

A salt and water are also produced when an acid reacts with a metal carbonate. The complete word equation is as follows:

metal carbonate + acid → salt + water + carbon dioxide

As with alkalis, the name of the metal in the metal carbonate becomes the first part of the name of the salt produced (see table 4.8).

name of carbonate	name of salt
sodium carbonate	sodium ....
potassium carbonate	potassium ....
calcium carbonate	calcium ....

Table 4.8 *Naming salts from metal carbonates*

Example 1

Sodium carbonate and hydrochloric acid react to produce sodium chloride, water and carbon dioxide.

Example 2

Potassium carbonate and sulphuric acid react to produce potassium sulphate, water and carbon dioxide.

Example 3

Calcium carbonate and nitric acid react to produce calcium nitrate, water and carbon dioxide.

Making salts from metal carbonates

Soluble metal carbonates, like sodium carbonate, produce alkaline solutions. They can be used to make salts by the method described earlier in figure 4.20.

For *insoluble* carbonates, like copper carbonate, a different method must be used. The insoluble carbonate is added to acid until the fizzing stops. At this point all of the acid has been neutralised. Any excess carbonate is filtered off and the salt solution is evaporated to obtain the salt. The technique is shown in figure 4.23.

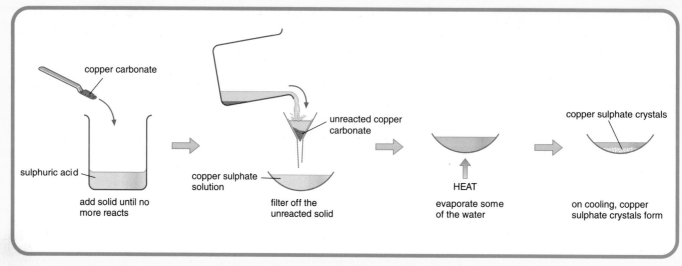

copper carbonate

sulphuric acid

add solid until no
more reacts

copper sulphate
solution

filter off the
unreacted solid

unreacted copper
carbonate

HEAT

evaporate some
of the water

copper sulphate crystals

on cooling, copper
sulphate crystals form

Figure 4.23 *Making copper sulphate crystals by neutralisation followed by evaporation*

Section Questions

15 Give the name of the salt that is formed by the reaction between each of the following:
a) calcium carbonate and hydrochloric acid,
b) sodium carbonate and nitric acid,
c) magnesium carbonate and sulphuric acid.

16 Write word equations for each of the following reactions:
a) sodium carbonate reacting with sulphuric acid,
b) zinc carbonate reacting with hydrochloric acid,
c) nickel carbonate reacting with nitric acid.

Everyday examples of neutralisation

Farmers and gardeners both use powdered chalk or limestone to neutralise soil acidity (see figure 4.24). We have already met one example of this on page 52 where sulphuric acid, which was sprayed on potatoes, is then neutralised with powdered limestone. Both chalk and limestone are mainly calcium carbonate, which can neutralise acids and raise the pH of soil.

Figure 4.24 *Gardeners use limestone (called 'lime' on the box) to neutralise soil acidity*

Calcium carbonate is also present in many indigestion remedies (see figure 4.25). All of these work by neutralising the excess stomach acid that causes the pain of acid indigestion.

We can show the neutralising effects of indigestion tablets by adding one to some dilute acid containing Universal indicator (see figure 4.26).

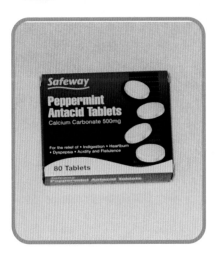

Figure 4.25 *Indigestion remedies work by neutralising excess stomach acid*

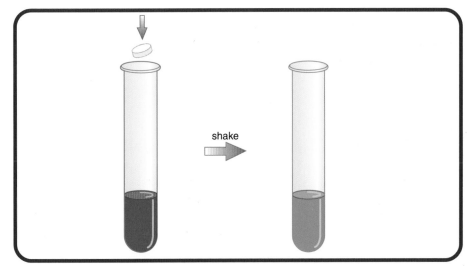

shake

Figure 4.26 *Adding an indigestion tablet to acid containing Universal indicator*

Large pieces of limestone are put into the streams where salmon spawn if the water there is too acidic. Lime, which is made by heating limestone, can also be used to neutralise acidity in water. It can be dropped from helicopters into inaccessible lochs and lakes, as was shown in figure 4.17 on page 48.

Section Questions

17 Why do farmers add chalk or limestone to soil?

18 What type of reaction takes place between limestone and the acids in an acidified lake or loch?

19 Give another example where calcium carbonate, or another compound, is used to counteract the effects of unwanted acidity.

Acid rain

Carbon and sulphur burn and combine with oxygen to form carbon dioxide and sulphur dioxide. Both of these gases dissolve in water to produce acid solutions. In the experiment shown in figure 4.27, samples of carbon and sulphur are heated in a bunsen flame to start them burning. They are then lowered into a gas jar containing oxygen, where they burn even better. The Universal indicator in the bottom of each gas jar changes colour, and shows that the gases are acidic.

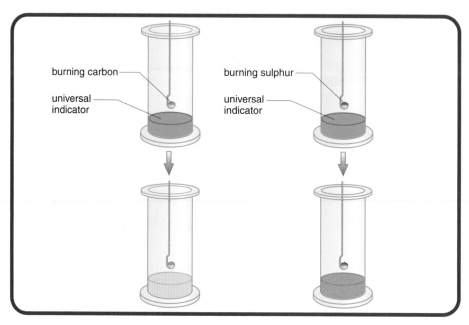

Figure 4.27 *Sulphur dioxide is more acidic than carbon dioxide*

Rain water is naturally slightly acidic because it contains dissolved carbon dioxide. This does not lower the pH very much and does not harm living things. Rain water containing only dissolved carbon dioxide is *not* considered to be 'acid rain'.

Sulphur is present as an impurity in fossil fuels, particularly in coal. Unfortunately, when these fuels are burned (see figure 4.28), the sulphur burns to produce sulphur dioxide. This dissolves to produce a stronger acid than carbon dioxide and so causes a lower pH. Rain water in which sulphur dioxide has dissolved has a much lower than normal pH and *is* described as **acid rain**.

Figure 4.28 *Coal-burning power stations, like this one on the Firth of Forth, produce thousands of tonnes of sulphur dioxide each year*

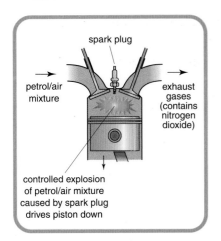

Figure 4.29 *A petrol engine produces nitrogen dioxide gas*

Another gas that contributes to acid rain is nitrogen dioxide. This is formed by reaction between nitrogen and oxygen in the air whenever a high enough temperature is reached. One of the places where there is a high enough temperature is near to the spark caused by the spark plugs inside a car's petrol engine (see figure 4.29).

Fortunately, not all of the nitrogen dioxide escapes to the air if the car is fitted with a catalytic converter. When hot, the catalysts in a catalytic converter can turn nitrogen dioxide back into harmless nitrogen.

In rain water, sulphur dioxide eventually produces sulphuric acid. Nitrogen dioxide produces nitric acid. In acid rain there is usually more sulphuric acid than nitric acid, but both are very harmful.

Section Questions

20 a) Name the gases produced when the following react with a plentiful supply of oxygen:
   i) carbon,
   ii) sulphur,
   iii) nitrogen.
b) When the gases produced in a) dissolve in water, is the solution acidic, alkaline or neutral?

What damage does acid rain cause?

Buildings

Acid rain has damaging effects on buildings made from carbonate rocks, such as limestone and marble. Both of these are made of calcium carbonate. Calcium carbonate is also a material that holds together grains of sand in sandstone. As acid rain wears away the calcium carbonate, the sandstone crumbles.

In Egypt, the ancient Sphinx is made mostly of limestone. Old photographs show that many of the features of the Sphinx have been eroded in recent years (see figure 4.30). Acid rain is thought to be the main cause of this.

Figure 4.30 *Many of the features of the Sphinx have been eroded away by acid rain*

In India, much of the Taj Mahal is made of marble, which has also been attacked by acid rain (see figure 4.31). One reason for this is that there is an oil refinery not far from it. This has released a lot of sulphur dioxide into the air over many years, causing rain in the area to be very acidic.

Figure 4.31 *Acid rain has caused a lot of damage to the Taj Mahal in recent years*

In Scotland, many of the older buildings are made of sandstone. Acid rain has caused much of the sandstone to crumble by attacking the calcium carbonate that holds the grains of sand together. There are examples in most towns, including the cathedral at Dunkeld (see figure 4.32).

Figure 4.32 *Much of the sandstone of which Dunkeld Cathedral is made has been eroded away over the centuries*

Iron and steel structures

Acid rain causes iron and steel structures to rust more rapidly than usual. Rail bridges have traditionally been made of steel (see figure 4.33), and have been considerably damaged by acids in rain water. The rail bridges over the rivers Forth and Tay were built over a hundred years ago, and quite a lot of the steel has rusted away.

Figure 4.33 *The Forth rail bridge contains 58 000 tonnes of steel*

Plant and animal life

Many lochs and lakes, in Scotland and elsewhere, have no fish in them as a result of acid rain. Some have pH values as low as 4. But it is not the acidity itself that kills the fish. It is more indirect. Because of the acid rain, aluminium ions are released from the surrounding soil. This causes the gills of the fish to become covered by mucus. The mucus prevents them from absorbing dissolved oxygen from the water. They die as a result.

Acid rain can be very damaging to trees. The acidity causes leaves and pine needles to turn yellow and to drop off. Eventually the trees die. Around the world, millions of trees have been damaged by acid rain (see figure 4.34).

Figure 4.34 *Trees damaged by acid rain*

The pH of soil can be lowered so much by acid rain that many crops cannot be grown. Fortunately, as has already been mentioned on page 54, neutralisers like chalk and limestone can be used to counteract the effects of acid rain in water and on land.

Section Questions

21 Which two gases are the main cause of acid rain?

22 Which acidic gas is produced when sulphur impurities in fossil fuels combine with oxygen in the air?

23 Which acidic gas is produced by the sparking of air in a car's petrol engine?

24 a) Name a type of rock that is damaged by acid rain.
   b) Give two other examples of things that are damaged by acid rain.

ACCESS 3 Subsection Test: Acids and alkalis

Part A

This part of the paper consists of four questions and is worth 4 marks.

1 What is the pH value of a neutral solution? (1)

   5 or 7

2 Which of the following is an example of a common household alkali? (1)

   baking soda or **vinegar**

3 Which of the following correctly describes the effect on the pH value of an acid as it is neutralised? (1)

   the pH moves down towards 7 or
   the pH moves up towards 7

4 What kind of solution is produced when sulphur dioxide dissolves in water? (1)

   an acidic solution or **an alkaline solution**

Part B

This part of the paper is worth 6 marks.

5 When an acid is diluted with water, what happens to the acidity? *les Acidic* (1)

6 Which acidic gas is produced when carbon reacts with oxygen? *CO₂* (1)

7 The table below shows pH values for some alkaline solutions.

alkali	pH
calcium hydroxide	12
ammonia	11
sodium hydroxide	14
lithium hydroxide	13

a) Present this information as a bar graph. (2)

b) Which of these substances is the weakest alkali? (1)

8 A sample of gas was tested with moist pH paper. The paper turned a blue-violet colour. Which of the following gases could have given this result? (1)

   A carbon dioxide

   B ammonia

   C nitrogen dioxide.

   Total 10 marks

Intermediate 1 Unit Test: Chemistry in Action

Part A

This part consists of twelve questions and is worth 12 marks.

In questions 1 to 6 choose the correct word to complete the sentences.

1 There are **more/less** metals than non-metals in the Periodic Table of elements. (1)

2 The gas that makes up about 80% of the air is **oxygen/nitrogen**. (1)

3 Most compounds with a name ending in '-ide' contain **three/two** elements. (1)

4 Large pieces of marble react **faster/slower** with acid than smaller pieces. (1)

5 The bonds inside molecules are **weaker/stronger** than those between molecules. (1)

6 Baking soda is an example of a common household **alkali/acid**. (1)

Questions 7 to 12 are multiple choice questions. Choose the correct letter.

7 Which of the following compounds is insoluble in water? You may wish to refer to page 4 of the data booklet. (1)
   A calcium nitrate
   B magnesium chloride
   C iron phosphate
   D sodium sulphate

8 Which of the following pairs of conditions is likely to produce the fastest speed of reaction between marble chips and hydrochloric acid? (1)

	chip size	acid concentration
A	small	high
B	small	low
C	large	high
D	large	low

9 When a metal carbonate reacts with an acid, what is the gas produced? (1)
   A hydrogen
   B oxygen
   C nitrogen
   D carbon dioxide

10 Ionic compounds are made up of: (1)
   A negatively charged ions only
   B positively charged ions only
   C negative and positive ions
   D atoms which have no charge

11 A part of the Periodic Table of elements is shown below. (1)

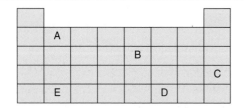

Which of the elements, A, B, C or D, has similar chemical properties to element E?

12 Chlorine is added to drinking water in order to: (1)
   A kill bacteria
   B strengthen teeth
   C remove lead compounds
   D make the water clear

Part B

This part consists of seven questions and is worth 18 marks.

13 The line graph below shows how the solubility of potassium chlorate changes with rise in temperature.

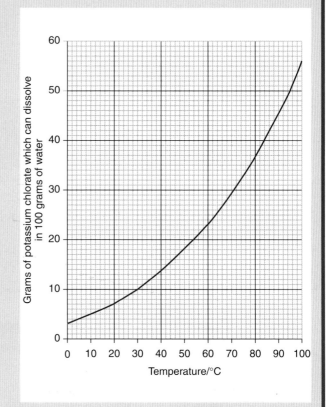

a) How many grams of potassium chlorate can dissolve in 100 grams of water at 30°C? (1)
b) At what temperature is the solubility of potassium chlorate 23 grams in 100 grams of water? (1)
c) What happens to the solubility of potassium chlorate as the temperature of the water increases? (1)

14 Metal sulphites react with acids to produce a salt, water and sulphur dioxide.

Potassium sulphite reacts with sulphuric acid to produce potassium sulphate, water and sulphur dioxide.

a) Write a word equation for this reaction. (1)

b) Write the chemical formula for sulphur dioxide. (1)

c) Name the salt produced when potassium sulphite reacts with hydrochloric acid. (1)

15 In the presence of a catalyst hydrogen peroxide releases oxygen gas.

Anne and Simon investigated the reaction using different catalysts.

Their experiments were the same, except that Anne used manganese dioxide as the catalyst and Simon used copper oxide. Anne's results produced graph A and Simon's produced graph B below.

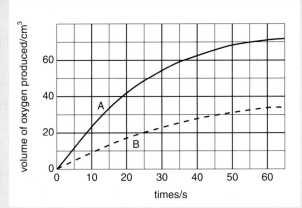

a) What is the effect of a catalyst on a chemical reaction? (1)

b) How long did it take to produce 30 cm³ of gas in each experiment? (1)

c) Which of the two oxides is the better catalyst? (1)

16 The table below shows the percentage by mass of some of the elements in the Earth's crust.

element	percentage
oxygen	47
silicon	28
aluminium	8
iron	5
sodium	3

a) Present this information as a bar graph. (2)

b) Which is the most plentiful *metal* in the crust of the Earth? (1)

17 The acid present in vinegar has the following molecular structure.

Write the chemical formula for this acid. (1)

18 In order to make copper sulphate crystals, Jack added copper carbonate to dilute sulphuric acid until no more reacted.

a) How should Jack remove the unreacted insoluble copper carbonate from the copper sulphate solution which he has made? (1)

b) Name all of the elements present in copper sulphate. (1)

19 Sulphur dioxide and nitrogen dioxide cause pollution in the atmosphere.

Sulphur dioxide is produced when fossil fuels are burned. Nitrogen dioxide is produced in car petrol engines.

a) Name a fossil fuel that produces sulphur dioxide when burned. (1)

b) What causes nitrogen dioxide to be produced in car petrol engines? (1)

c) Name two things that are harmed as a result of sulphur dioxide and nitrogen dioxide dissolving in rain water. (1)

Total 30 marks

Unit 1 Glossary of Terms

acid   A substance that gives a solution with a pH of less than 7.

acid rain   Rain that is more acidic than usual due to dissolved sulphur dioxide and nitrogen dioxide.

air   A mixture of gases, approximately 80% nitrogen and 20% oxygen.

alkali   A substance that gives a solution with a pH of more than 7.

alkali metals   Reactive metal elements in column 1 of the Periodic Table.

atom   The smallest part of an element that can exist.

atomic number   A special number given to each element in the Periodic Table.

catalyst   A substance that speeds up a reaction but is not used up by the reaction.

chemical reaction   A chemical process in which one or more new substances are formed (usually accompanied by an energy change and a change in appearance).

chemical symbol   One or two letters used to represent an element, for example C for carbon, Al for aluminium.

compound   A substance in which two or more elements are joined together chemically. Compounds are formed when elements react together.

element   The simplest kind of substance that cannot be broken down into anything simpler.

enzyme   A catalyst that affects living things.

halogens   Reactive non-metal elements in column 7 of the Periodic Table.

ion   An atom or group of atoms that possess a positive or negative charge.

metals   Shiny elements that conduct electricity.

mixture   A mixture is formed when two or more substances come together but do not react, e.g. air and sea water.

molecule   A group of two or more atoms held together by strong bonds.

neutral solution   One with a pH of 7.

neutralisation reaction   A reaction of acids with alkalis or metal carbonates that moves the pH of the solution towards 7.

noble gases   Unreactive gases in column 0 of the Periodic Table.

non-metals   Elements that do not conduct electricity (carbon is an exception). Mostly not shiny.

Periodic Table   An arrangement of the elements in order of increasing atomic number. Chemically similar elements occur in the same vertical column.

pH   A number that indicates the degree of acidity or alkalinity of a solution. The pH scale ranges from below 0 (very acidic) to above 14 (very alkaline).

salt   A compound formed by the result of a reaction between an acid and an alkali or metal carbonate, e.g. sodium chloride.

saturated solution   One in which no more substance can be dissolved.

solution   A liquid with a substance dissolved in it.

symbol   *See* chemical symbol.

toxic   Poisonous.

word equation   An equation that gives the names of reactants and products, e.g.

carbon + oxygen → carbon dioxide

Unit 2

Everyday Chemistry

The topics covered in this unit are
Metals
Personal needs
Fuels
Plastics

5 Metals

Extracting metals

Can you think of any metal elements that are found in the crust of the Earth? Possibly you know that gold is one of these metals. Gold can still be found in streams and rivers in Scotland (see figure 5.1). Scottish kings were once crowned with Scottish gold, and gold has long been used for ornaments (see figure 5.2).

Figure 5.1 *Panning for gold*

Gold is an unreactive metal, and it is found uncombined because it does not react readily with other elements such as oxygen. Other unreactive metals that are found uncombined in the Earth's crust include silver and copper.

Most metals are found combined with other elements in **ores**. Ores are naturally occurring compounds from which metals are extracted. Some common ores are given in table 5.1 (and see figures 5.3 to 5.5).

ore	compound present
bauxite	aluminium oxide
haematite	iron oxide
galena	lead sulphide
malachite	copper carbonate

Table 5.1 *Some metal ores*

Figure 5.2 *A gold ornament made thousands of years ago*

Figure 5.3 *Malachite*

Figure 5.4 *Haematite*

Figure 5.5 *Galena*

Section Questions

1 Name three metals that are found uncombined in the Earth's crust.

2 Why are so few metals found uncombined in the Earth's crust?

Extracting metals from ores

Copper can be extracted from a copper ore, such as malachite, by heating the ore strongly with carbon in a test tube (see figure 5.6). The test tube is then plunged into cold water in a beaker (figure 5.7). The test tube shatters and copper can be found at the bottom of the beaker. The water prevents the hot copper from coming into contact with air. When hot copper reacts with oxygen in the air, it forms copper oxide.

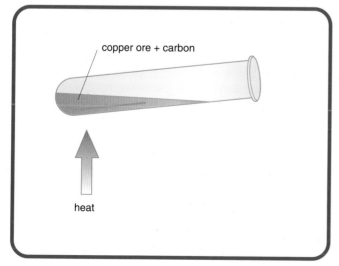

Figure 5.6 *Heating copper ore with carbon*

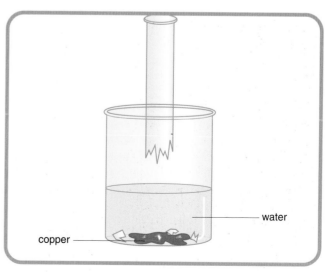

Figure 5.7 *Separating the extracted copper*

Extracting iron in a blast furnace

Like copper, iron and some other metals can be extracted from their ores by heating with carbon. On an industrial scale, the extraction of iron takes place in a blast furnace (see figure 5.8).

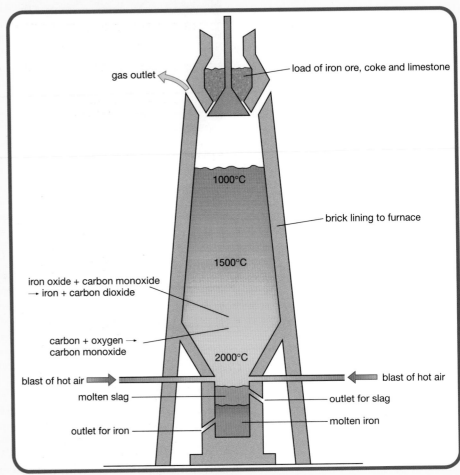

Figure 5.8 *Inside a blast furnace*

Blast furnaces can operate for two years non-stop. This is how they work:
- Iron ore, coke and limestone are fed in at the top of the furnace.
- A blast of hot air is forced up through the furnace.
- Near the bottom of the furnace, carbon in the coke reacts with oxygen to form carbon monoxide gas.
- As it rises up through the furnace, carbon monoxide reacts with iron oxide in the iron ore to produce iron and carbon dioxide.
- It is so hot in the furnace that the iron formed is molten and falls to the bottom of the furnace.
- The limestone that was added reacts with impurities to form a light 'slag', which floats on top of the molten iron.
- Molten iron and slag are run off separately from the bottom of the blast furnace.

The slag is useful and can be used for road building. A modern blast furnace can produce about 10 000 tonnes of iron a day. More iron is produced in the world than any other metal.

Extracting aluminium using electricity

Many metals that are more reactive than iron are extracted using electricity. Aluminium is one such metal. The process to extract aluminium takes place in steel tanks lined with carbon (see figure 5.9).

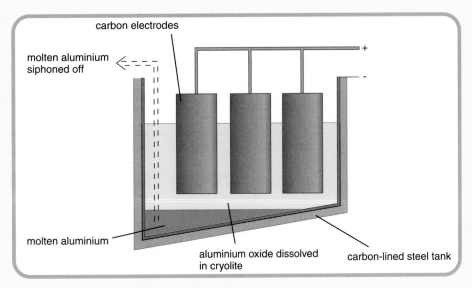

Figure 5.9 *A cell used for the extraction of aluminium*

- Reactive metals, like aluminium, are extracted by passing a direct current through a molten compound.
- In the case of aluminium, aluminium oxide is dissolved in molten cryolite, which is another aluminium compound.
- Positive aluminium ions are attracted to the negatively charged carbon lining of the cell. They turn into aluminium atoms when they reach the carbon lining.
- The aluminium forms a liquid layer at the bottom of the cell.
- From time to time the molten aluminium is siphoned off.

A typical cell can produce more than a tonne of aluminium in one day. The world production of aluminium is second only to that of iron.

Extracting copper using electricity

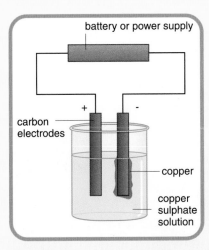

Figure 5.10 *A cell used to extract copper from copper sulphate solution*

In the laboratory we can extract copper by passing a direct current through a solution of one of its compounds (see figure 5.10).

- Unreactive metals, like copper, can be extracted by passing a direct current through a solution of one of their compounds.
- Positive copper ions are attracted to the negative electrode. They turn into copper atoms when they reach the electrode.
- The copper is seen as a pinkish brown deposit on the electrode.

Section Questions

3 During the extraction of zinc, zinc oxide reacts with carbon to produce zinc and carbon dioxide. Write a word equation for this reaction.

4 What material are the electrodes made of for the extraction of aluminium using electricity?

5 At which electrode are metals obtained when a direct current is being used to extract them?

Properties of metals

We are surrounded by metals. Cars, lorries, buses, ships, bridges, trains, railway lines and many more things are made using metals. We can use metals for so many different jobs because of their useful properties.

The following are some of the most important **properties** of metals.

Strength

Metals that are strong are used to make car bodies, bicycle frames, ships' hulls, etc. (see figure 5.11).

Malleability

Metals are malleable. This means that they can be shaped by hammering or rolling and can be bent without breaking (see figure 5.12).

Figure 5.11 *The steel hull of this ferry must be very strong*

Figure 5.12 *The malleability of aluminium allowed it to be shaped to form the body of this car*

Conduction of electricity

All metal elements conduct electricity. Non-metals, with the exception of graphite, do not. Copper wires are used in household cables and flexes for electrical appliances (see figure 5.13). Aluminium cables carry the current in the National Grid (see figure 5.14).

Figure 5.13 *Copper wires connect this kettle to a socket*

Figure 5.14 *Aluminium is used to carry electricity across country*

Conduction of heat

Metals are good conductors of heat. We use metals like aluminium, iron and copper for cooking pots and pans (see figure 5.15). They are able to conduct the heat from the cooker to the food inside the pan.

Figure 5.15 *Iron cooking pots are good at conducting heat to the food inside*

Density

This the mass of a substance in a given volume. A high-density material is much heavier than the same volume of a low-density material. Aluminium is a metal with a low density. It is used to build aircraft (see figure 5.16). Lead is a metal with a high density. It is attached to divers' suits (see figure 5.17).

Figure 5.16 *Aircraft are made of aluminium because of its low density*

Figure 5.17 *Lead weights are attached to divers' suits because of its high density*

Section Questions

6 Use page 5 of the data booklet to help you to answer these questions.
a) Which metal is strong and has a low density?
b) Which metal has a high density and is a liquid at room temperature?
c) Give two uses for tin.

Alloys

The properties of metals can be changed and improved by making **alloys**. Most alloys are mixtures of two or more metals, but in a few cases non-metals are also added. The usual way to make an alloy is to melt together the elements that make it up.

Pure aluminium is too soft for use as a car body or for aircraft manufacture. Mixing it with magnesium to make an alloy makes it stronger. It is this alloy that is used to make cars and aircraft. The alloy of aluminium and magnesium is called **magnalium**. There are many alloys in common everyday use.

Steels are alloys in which the main metal element is iron. **Mild steel** is an alloy of iron and the non-metal carbon. It is harder and stronger than pure iron. Unfortunately, mild steel rusts easily. **Stainless steel** is much more rust-resistant (see figure 5.18). It is an alloy of iron, chromium and nickel.

Figure 5.18 *This stainless steel cutlery is guaranteed against rusting for thirty years*

Figure 5.19 *The connections in this plug are made of brass*

Brass is harder than the copper and zinc that make it up. It is widely used for letter boxes, door handles, ornaments and electrical contacts in plugs (see figure 5.19).

Solder is an alloy of tin and lead. It is more easily melted than either tin or lead individually. Because of its low melting point, it is widely used for making electrical connections (see figure 5.20).

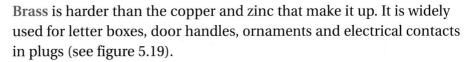

Figure 5.20 *Soldered connections are made of tin and lead*

Figure 5.21 *Our coins are made of alloys*

There are many other alloys. **Bronze**, which is used for statues, is made of copper and tin. Our coins are also alloys (see figure 5.21), because they need to be harder wearing than pure metals. Their compositions are given in table 5.2.

coin type	values	made from ...
'copper'	1p, 2p	copper-coated steel
'silver'	5p, 10p, 20p, 50p	copper and nickel
'gold'	£1, £2	copper, nickel and zinc

Table 5.2 *What our coins are made of*

Summary

The compositions of the alloys we have mentioned are given in table 5.3.

alloy	made from ...
mild steel	iron and carbon
stainless steel	iron, chromium and nickel
brass	copper and zinc
solder	lead and tin
bronze	copper and tin

Table 5.3 *Composition of some alloys*

Section Questions

7 Using information on page 3 of the data booklet, name a metal that is more dense than mercury.

8 Wood's metal is used as an easily melted plug in shop sprinkler systems. It contains bismuth (50%), lead (25%), cadmium (12.5%) and tin (12.5%).
a) What is meant by the term 'an alloy'?
b) Draw a bar graph to show the composition of Wood's metal.

Prescribed Practical Activity

Electrical conductivity

Information

The aim of the experiment is to test the electrical conductivity of some metals and non-metals. From the results, a general rule can be worked out about the electrical conductivity of elements.

What to do

1 Set up a circuit like the one shown in figure 5.22.

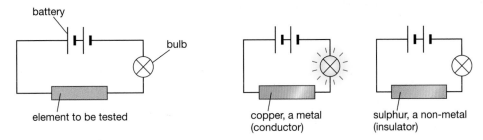

Figure 5.22 *Testing elements for electrical conduction*

2 Take one of the elements provided and test its electrical conductivity.

3 Record the following information in a table:
- the name of the element,
- whether it is a metal or non-metal,
- whether it conducts or not.

4 Use the data booklet to find whether the element is a metal or a non-metal.

5 Repeat the experiment for the other elements provided.

6 Record your results.

Results table

Some extra elements are shown in **bold** in the table, which, for safety reasons, you have not been asked to work with. Whether they conduct or not is given, but you should find out whether they are metals or non-metals by using the data booklet.

element	metal/non-metal	conductor/non-conductor
aluminium	metal	conductor
carbon	non-metal	conductor
copper	metal	conductor
iron	metal	conductor
nickel	metal	conductor
sulphur	non-metal	non-conductor
zinc	metal	conductor
iodine	non-metal	**non-conductor**
phosphorus	non-metal	**non-conductor**
mercury	metal	**conductor**
bromine	non-metal	**non-conductor**
selenium	non-metal	**non-conductor**

Conclusion

- Metals conduct electricity but non-metals do not.
- The non-metal carbon is an exception to the rule as it is a conductor of electricity.

Reactions of metals

Some metals react quickly and easily with other substances. A freshly cut piece of sodium has a shiny, silvery surface (see figure 5.23). However, within seconds the surface goes dull as sodium reacts rapidly with gases in the air (see figure 5.24). Sodium is a reactive metal.

Figure 5.23 *Freshly cut sodium is shiny …*

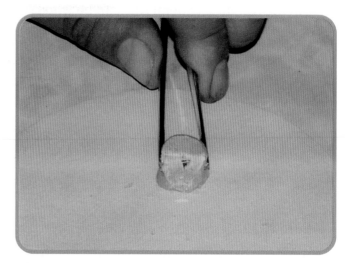

Figure 5.24 *… but it soon turns dull*

Copper stays shiny (see figure 5.25) much longer than sodium before the surface becomes dull. It can take years for a green coating (see figure 5.26) to form on the roof of a building that is covered in copper.

Figure 5.25 *Shiny copper sheet …*

Figure 5.26 *… eventually turns green*

Gold is not affected by gases in the air. Gold on the dome of a mosque or temple stays shiny for many years (see figure 5.27). Gold is a very unreactive metal.

Figure 5.27 *Why has this gold-covered temple stayed shiny for so long?*

From the previous observations, we can put the three metals, sodium, copper and gold, into an order of reactivity. Sodium is more reactive than copper. Copper is more reactive than gold.

A **reactivity series of metals** is a kind of 'league table', which puts them in order of reactivity. Table 5.4 shows a reactivity series for some common metals (which is also given on page 6 of the data booklet). The most reactive metals are at the top and the least reactive are at the bottom.

metal	reactivity
potassium	most reactive
sodium	
lithium	
calcium	
magnesium	
aluminium	
zinc	
iron	
tin	
lead	
copper	
mercury	
silver	
gold	least reactive

Table 5.4 *A reactivity series of metals*

Metals can be placed in a reactivity series by observing their reactions with, for example, oxygen, water and dilute acids.

Reaction with oxygen

All of the metals above mercury in the reactivity series combine with oxygen when heated (see figure 5.28). This produces a metal oxide. The higher a metal is in the series, the more violent the reaction between the metal and oxygen. Magnesium, for example, burns fiercely with a bright white light.

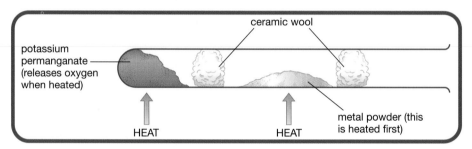

Figure 5.28 *Metals glow as they react with oxygen*

The apparatus shown in figure 5.28 can be used to study the reactivities of the metals from magnesium down to copper in the reactivity series. The general word equation is:

metal + oxygen → metal oxide

For example:

magnesium + oxygen → magnesium oxide

The metals above magnesium in the reactivity series are too reactive to risk in this experiment. Potassium, sodium and lithium are stored in oil. This keeps them out of contact with oxygen and other reactive gases in the atmosphere.

Reaction with water

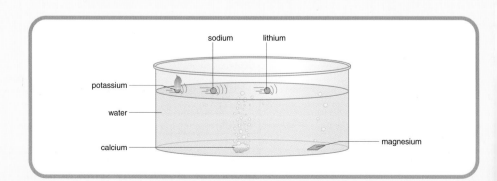

Figure 5.29 *Metals reacting with water*

The metals above aluminium in the reactivity series react with water (see figure 5.29). The products are the metal hydroxide and hydrogen gas. Pieces of potassium, sodium and lithium move around on the surface of the water as they react. They do this because they are less dense than water. In the case of potassium, the hydrogen released catches fire.

Using calcium, the hydrogen given off can be collected quickly and quite safely using test tubes as in figure 5.30.

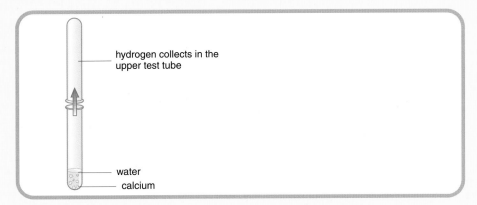

hydrogen collects in the upper test tube

water
calcium

Figure 5.30 *Hydrogen can be collected when calcium reacts with water*

When metals react with water, a metal hydroxide is formed in addition to hydrogen. The general word equation is:

metal + water → metal hydroxide + hydrogen

For example:

sodium + water → sodium hydroxide + hydrogen

Testing for hydrogen

A lit taper is applied to a test tube containing a mixture of hydrogen and air. The result is that the hydrogen burns with a 'pop' (see figure 5.31). We usually say that **hydrogen burns with a pop**.

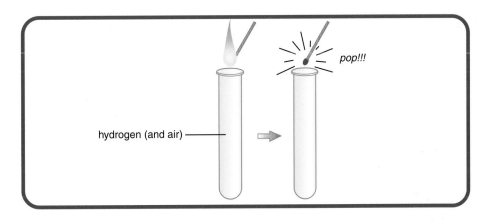

pop!!!

hydrogen (and air)

Figure 5.31 *Hydrogen burns with a pop*

Reaction with dilute acids

All of the metals above copper in the reactivity series react with dilute acids. The products are a salt and hydrogen. The general word equation is:

metal + acid → salt + hydrogen

For example:

zinc + hydrochloric acid → zinc chloride + hydrogen

When a metal reacts with a dilute acid, it produces bubbles of hydrogen gas. The faster the bubbles are produced, the more reactive the metal (see figure 5.32).

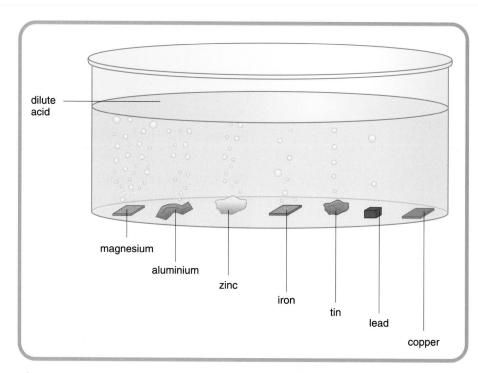

Figure 5.32 *Metals reacting with dilute acid*

Summary

The reactions of metals with oxygen, water and dilute acid are summarised in table 5.5. The summary also appears on page **6** of the data booklet.

metal	reaction with		
	oxygen	water	dilute acid
potassium	metals which react with oxygen	metals which react with water	metals which react with dilute acid
sodium			
lithium			
calcium			
magnesium			
aluminium		no reaction	
zinc			
iron			
tin			
lead			
copper			no reaction
mercury	no reaction		
silver			
gold			

Table 5.5 *Reactions of metals*

9 When copper is heated, it reacts with oxygen and turns black.
   a) Name the compound formed.
   b) Write a word equation for the reaction.

10 Lithium is stored in oil. What does this tell you about its reactivity?

11 Describe the test for hydrogen.

12 In an experiment, zinc and tin were both added to a little dilute acid in separate test tubes. What observation would show that zinc is more reactive than tin?

Reaction of metals with acid

Information

The aim of the experiment is to place the three metals, zinc, magnesium and copper, in order of reactivity. This is done by watching how quickly they react with hydrochloric acid.

What to do

1 Add some dilute hydrochloric acid to three test tubes.
2 Add a piece of zinc to one test tube (see figure 5.33).
3 Add a piece of magnesium to the second test tube.

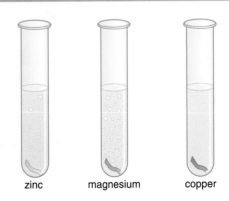

zinc   magnesium   copper

Figure 5.33 *Reactions of metals with dilute hydrochloric acid*

4 Add a piece of copper to the third test tube.

5 For each experiment record the following in a table:
- the name of each metal,
- whether bubbles are produced or not,
- the speed of the reaction.

Results table

metal	bubbles produced?	reaction speed
zinc	yes	slow
magnesium	yes	fast
copper	no	no reaction

Conclusion

The order of reactivity of the three metals is as follows.

magnesium (most reactive)

zinc

copper (least reactive)

Corrosion

If you have an old bicycle, you may have seen how corrosion can affect metals. For example, the steel in the cycle chain can slowly turn to rust, making it weaker and weaker until it eventually snaps. Motor cars can fail their MOT test if they are too corroded (see figure 5.34). Corrosion weakens the structure of cars and can make them unsafe.

Figure 5.34 *Would this car pass its MOT test?*

During corrosion of a metal, a chemical reaction takes place. The metal element is changed into a compound. In the case of iron (see figure 5.35), the reddish brown compound iron oxide (see figure 5.36) is produced.

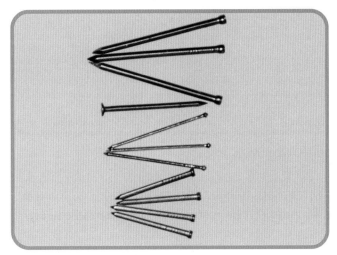

Figure 5.35 *Iron is shiny …*

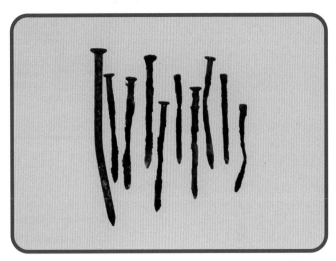

Figure 5.36 *… until it rusts*

Different metals corrode at different rates. Copper, for example, corrodes more slowly than iron, forming the green compound copper carbonate on its surface. Look back at figures 5.25 and 5.26 on page 76 to see this.

The rusting of iron

Iron is the most widely used metal. Much of the iron produced is to replace that which has been lost due to **rusting**. Rusting is the term used for the corrosion of iron. Many metals corrode, but only one (iron) rusts.

Rusting is a major problem. It costs millions of pounds each year to replace rusty cars, lorries, ships, etc. It is therefore important to find out what causes rusting and to try to stop it.

Three identical iron nails are placed in test tubes as shown in figure 5.37. In tube 1, calcium chloride is added to remove any moisture from the air. In tube 2, the water is boiled to drive out dissolved air from the water. The layer of oil stops any air dissolving back into the water. After a few days, the nails are examined to see if they have rusted. The results of the experiment are shown in figure 5.37.

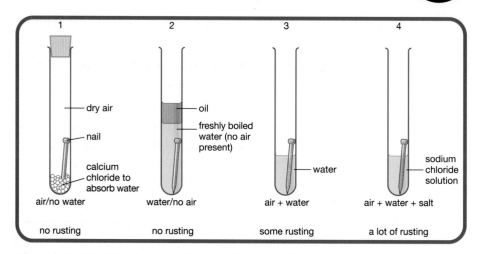

Figure 5.37 *What causes iron nails to rust?*

Evidently both air *and* water are needed for rusting to occur. Rusting therefore requires air, but could it be just one of the gases in the air that is needed?

The experiment in figure 5.38 shows that *oxygen* is needed for rusting to take place. As the moist iron filings in the test tube rust, the water level in the test tube rises. It rises about one-fifth of the way up the inside of the test tube. Do you remember that oxygen makes up about one-fifth of the air. Note that a second empty test tube is used as a control – to make sure that the test is fair.

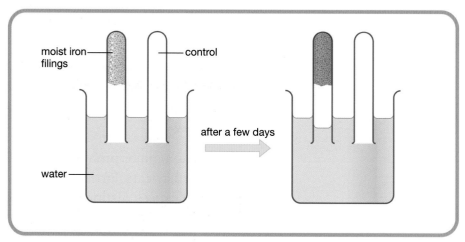

Figure 5.38 *Which gas is used up during rusting?*

We can now conclude that, for iron to rust, *oxygen* and *water* must be present.

A rust indicator

A good way of detecting even the smallest amount of rusting is to use **rust indicator**. This is a pale yellow solution that turns blue if rusting is taking place. The amount of blue colour produced also shows how much rusting has taken place. The experiment shown in figure 5.39 tells us that acid rain and salt water both speed up the rusting of iron. The colours develop quite quickly.

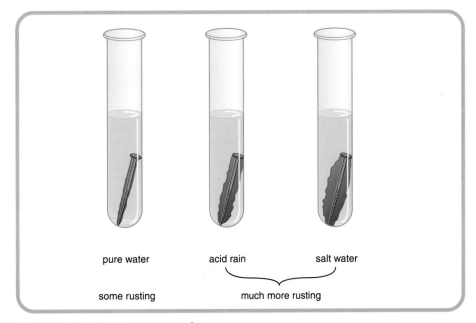

pure water　　　　acid rain　　　　salt water

some rusting　　　　　much more rusting

Figure 5.39 *Using rust indicator*

Salt spread on roads in winter increases the rate of corrosion on car bodywork (see figure 5.40).

Figure 5.40 *Keeping roads ice-free causes more rust on cars*

Preventing corrosion

For a metal like iron to corrode, it must come into contact with oxygen and water. One way to prevent the corrosion is therefore to put a barrier round the metal to keep out oxygen and water (see figure 5.41).

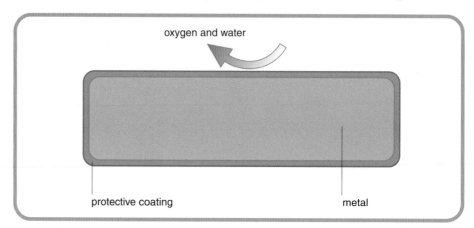

Figure 5.41 *Preventing corrosion – the barrier method*

Paint is one of the most common forms of physical barrier. The Forth Railway Bridge (see figure 5.42) is protected using paint, as are cars, trains, lorries, ships, etc.

Figure 5.42 *The Forth Railway Bridge – painting this takes a long time!*

Figure 5.43 *Without its flexible coating of oil, this chain would soon rust*

Some barriers must be very flexible. The moving parts of machines, such as a bicycle chain, are protected using **oil or grease** (see figure 5.43). Metals can also be given a protective flexible coating by coating them with **plastic**. Steel fencing lasts much longer if it is plastic coated.

Other metals, such as zinc, tin and chromium, can be used to provide a protective layer on iron and steel. All three metals can be deposited as a thin layer by means of a process called **electroplating**. In the laboratory, an iron nail can be given a coating of zinc as shown in figure 5.44. The nail is made the negative electrode in a solution containing positive zinc ions. A piece of zinc is used as the positive electrode. Positive zinc ions are attracted to the negative iron nail. There they gain electrons from a dc supply, such as a battery or power supply, and form a layer of zinc on the nail.

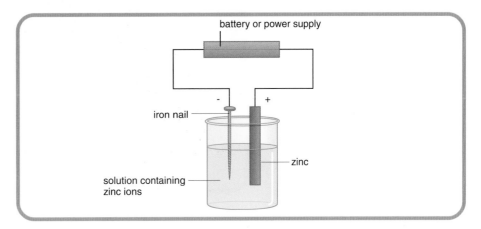

Figure 5.44 *Electroplating an iron nail with zinc*

A very effective way of protecting an iron or steel object from corrosion is by dipping it in molten zinc. This process is called **galvanising**. The car manufacturers Jaguar dip the entire body shell of their cars in a huge bath containing molten zinc (see figure 5.45).

Figure 5.45 *A jaguar car being dipped in molten zinc for protection*

Figure 5.46 *The Forth Road Bridge*

If an object is too large to dip in molten zinc, it can be sprayed instead. The steel used to make the Forth Road Bridge (see figure 5.46) was sprayed with 500 tonnes of molten zinc.

Many food manufacturers sell their products in cans or 'tins'. Some of these are steel cans that are coated (or 'plated') with tin (see figure 5.47). Tin is a fairly unreactive metal. It is shiny and gives good protection for the steel.

Figure 5.47 *All of these 'tins' are made of tin-coated steel*

Galvanising and tin-plating compared

What do you think will happen when the coatings on galvanised steel and tin-plated steel are deeply scratched? You might expect the steel to rust. We can find out by putting drops of yellow rust indicator on scratches on these surfaces. The results of the experiment are shown in figure 5.48.

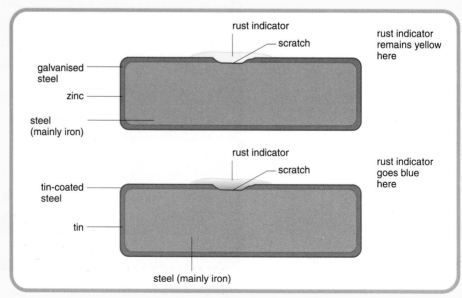

Figure 5.48 *The results of a corrosion experiment*

The zinc-coated (galvanised) steel has *not* corroded. There is no evidence of rusting. The yellow rust indicator stays yellow. The tin-coated steel, however, *does* corrode and rusts as shown by the blue colour of the indicator.

The reason why zinc protects the iron in steel is because zinc is *higher* in the reactivity series than iron. Tin is *below* iron in the series, and therefore does not protect the iron from corrosion.

Another corrosion experiment!

A more reactive metal has only to be attached to iron in order to prevent it from rusting. In the experiment shown in figure 5.49, magnesium and copper are attached to iron nails. These are then placed in rust indicator.

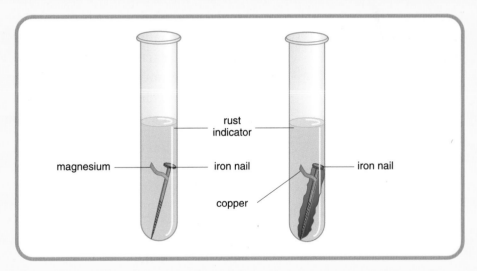

Figure 5.49 *Another corrosion experiment!*

The iron nail attached to magnesium does not rust. This is because magnesium is more reactive than iron. Copper does not prevent the rusting of iron. This is because copper is less reactive than iron.

Bags of scrap magnesium are attached to iron pipelines in order to stop the iron from rusting.

Anodising

Aluminium is higher than iron in the reactivity series. Without protection, iron corrodes rapidly. Why, then, can aluminium metal be used without any protective coating being applied? Have you ever seen a *painted* aluminium greenhouse frame (see figure 5.50), for example?

Figure 5.50 *The aluminium frame of this greenhouse resists corrosion – why?*

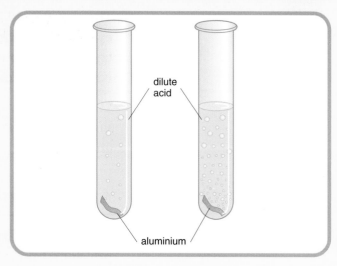

Figure 5.51 *The oxide coating has been removed from the piece of aluminium on the right*

Unlike iron, aluminium is protected by a very thin, but very protective, coating of aluminium oxide. With the coating intact, aluminium reacts only very slowly with a dilute acid. However, if the coating is removed, the reaction is much faster (see figure 5.51). The aluminium oxide coating is easily removed, either using steel wool, or by dipping in copper chloride solution.

The aluminium oxide coating on aluminium can be made thicker by a process called **anodising**. The thicker coating is even more protective *and* it can be dyed as well.

Anodising involves passing a direct current through dilute sulphuric acid. The aluminium object to be anodised is used as the positive electrode. The negative electrode can also be made of aluminium. The apparatus for anodising is shown in figure 5.52.

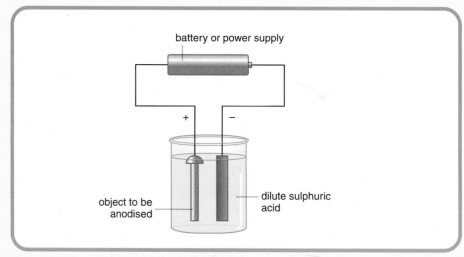

Figure 5.52 *Anodising an aluminium object*

Section Questions

13 What is the common name for the corrosion of iron?

14 Which two substances are needed for the corrosion of iron?

15 What effect do acid rain and salt have on the rate of corrosion?

16 Give two examples of materials that are used as a barrier to prevent corrosion.

17 Which metal is used to coat iron during galvanising?

18 What effect does anodising have on the oxide layer that is present on aluminium?

Batteries

How many things can you think of that are 'battery-operated'? Computer games, mobile phones, remote controls, torches, toys, radios, miniature televisions, cameras, watches and clocks are only a few (see figure 5.53).

Figure 5.53 *All of these are battery-operated*

Figure 5.54 *Batteries come in all shapes and sizes*

Batteries can be large or small (see figure 5.54). They can be capable of producing a large current for a short time, like a car battery. They can also be very small and produce a tiny current for a long time, like the battery in a watch. Some are rechargeable and some are not.

An example of a rechargeable battery is the lead–acid battery (see figure 5.55) used to start a car. Nickel–cadmium batteries (see figure 5.56) are also rechargeable, and can be used in a radio for example.

Figure 5.55 *A lead–acid battery*

Figure 5.56 *Nickel–cadmium battery*

In all batteries, electricity comes from a chemical reaction. We can show an example of this by joining a bulb to pieces of copper and zinc. When these metals are placed in dilute sulphuric acid, the bulb lights up (see figure 5.57). The chemical reaction producing the electricity can be seen by the bubbles of gas given off.

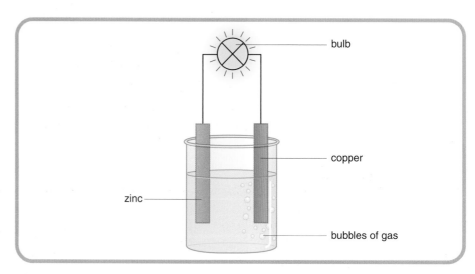

Figure 5.57 *A chemical reaction producing electricity*

Ordinary batteries need to be replaced once the chemicals inside them are used up. Rechargeable batteries cost more but the chemical reactions taking place inside them can be reversed. This means that the chemicals needed can be remade in the process of recharging. Some batteries can be recharged hundreds of times.

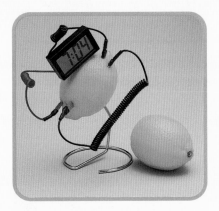

Figure 5.58 A 'lemon' clock

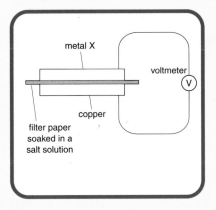

Figure 5.59 A cell using two metals

More about batteries

Electricity can be produced by connecting two different metals to form a cell. A solution containing ions is needed to complete the circuit between the two metals. The makers of the 'lemon' clock in figure 5.58 use this information.

Two strips of metal, one copper and the other zinc, are pushed into a lemon. A digital clock is then attached to the metal strips. The lemon contains ions, which complete the circuit. Similar clocks can be made using other fruits and even vegetables, such as potatoes.

Another league table of metals

In place of the fruit or vegetable in the 'lemon' cell, we can use a piece of filter paper. This has to be soaked in a salt solution to provide ions. Metals are placed on either side of the filter paper and connected by a voltmeter. In the experiment, one metal, such as copper, is kept the same. The other metal (metal 'X') is changed (see figure 5.59).

A typical table of results is shown in table 5.6. The metals magnesium, zinc, iron and lead were used, in turn, as metal 'X'.

metal 'X'	voltage
magnesium	2.1
zinc	1.0
iron	0.7
lead	0.3

Table 5.6 Voltages from simple cells using copper as one of the metals

From table 5.6 we can see that different pairs of metals produce different voltages. The voltage is also related to the positions of metals in the reactivity series. The further apart the metals are in the reactivity series, the higher the voltage.

Section Questions

19 Complete the following sentence:

In a battery, electricity comes from a chemical
....

20 Give two examples of rechargeable batteries.

21 A cell to produce electricity can be made using two metals and a solution containing ions. What is the purpose of this solution?

22 Two cells are made, one containing zinc and copper, the other containing magnesium and copper. With the help of page 6 of the data booklet, decide which cell will produce the greater voltage.

Everyday Chemistry

ACCESS 3 Subsection Test: Metals

Part A

This part of the paper consists of four questions and is worth 4 marks.

1 Which of these two metals is found uncombined in the Earth's crust? (1)

zinc or **silver**

2 Which of the following is an alloy? (1)

brass or **copper**

3 Which gas is produced by the reaction of metals with acid? (1)

oxygen or **hydrogen**

4 What does the surface of a metal turn into when it corrodes? (1)

a mixture or **a compound**

Part B

This part of the paper is worth 6 marks.

5 Complete the following word equation: (1)

iron + ... → iron oxide

6 A sample of gas was found to burn with a pop. Which of the following could have given this result? (1)

A nitrogen
B hydrogen
C oxygen

7 Why does painting steel protect the steel from corrosion? (1)

8 The table shows voltages from some single cell batteries.

battery	voltage
zinc–carbon	1.5
lithium	3.0
lead–acid	2.0
zinc–copper	1.0

a) Present this information as a bar graph. (2)

b) Which of these batteries is rechargeable? (1)

Total 10 marks

Personal needs

Keeping clean

How clean could you get your skin or hair simply by using just water? Would you wash a car mechanic's overalls, covered in oil and grease, in water alone? Our skin and hair also become greasy, and that is the problem. Fortunately, cleaning chemicals in the form of **soaps** and **detergents** are available. These can break up oil and grease into tiny droplets, which then mix with water and can be washed away.

We can make oil and water mix by shaking. However, the mixture soon separates into two layers again on standing (see figure 6.1).

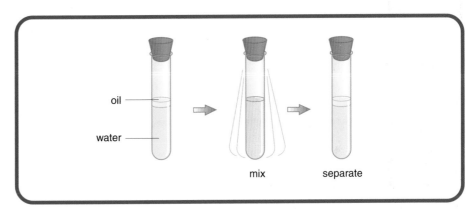

Figure 6.1 *Oil and water mix on shaking – but then separate*

If we add a few drops of detergent to the oil and water, something different happens. On shaking, the oil and water mix, but they *do not* separate on standing (see figure 6.2).

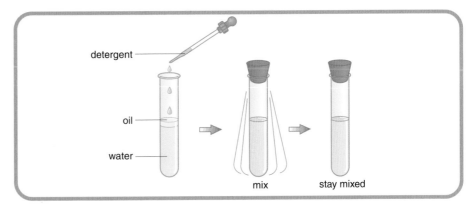

Figure 6.2 *When shaken with a little detergent, oil and water mix and stay mixed*

The effect of a soap solution on a mixture of oil and water is just the same. Soaps and detergents are similar. Some are molecular compounds, others are ionic compounds, but the part that cleans is much the same. In figure 6.3 you will see that this special part is like a tadpole in shape. It has a head and a tail. The clever bit is that the head is soluble in *water* and the tail is soluble in *oil*.

water-soluble head

oil soluble tail

Figure 6.3 *A soap or detergent 'tadpole'*

When detergent or soap is added to oil and water, the tail goes into the oil while the head stays in the water (see figure 6.4).

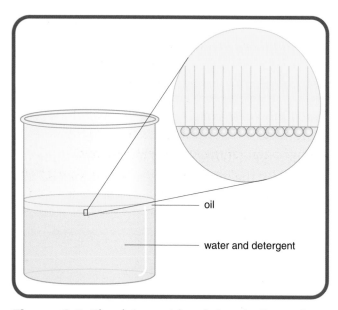

oil

water and detergent

Figure 6.4 *The detergent head stays in the water, with the tail in the oil*

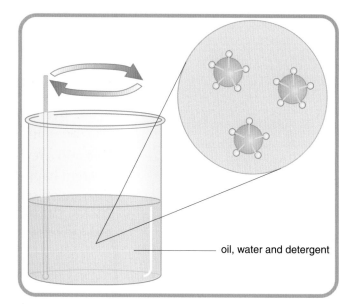

oil, water and detergent

Figure 6.5 *Stirring breaks up the oil into droplets – which stay broken up*

When the mixture is agitated, the oil breaks up into droplets. The oil does not separate from the water again because each oil droplet is surrounded by detergent 'tadpoles' (see figure 6.5).

There are many examples of products that contain cleaning chemicals such as soaps and detergents. These include not only soaps and detergents, but also shampoos, washing-up liquids and dish-washing powders. All of the products in figure 6.6 contain cleaning chemicals.

Figure 6.6 *All of these products contain cleaning chemicals*

Hard and soft water

Do you live in an area of **hard water**? What evidence is there if the water is 'hard'? If the area has hard water, the elements on electric kettles become 'furred up'. Also, when you use soap in hard water, a scum is formed. If no scum forms, the water is said to be 'soft'.

Using soaps for cleaning in areas of hard water is wasteful. This is because soaps cannot perform their cleaning duties until the 'hardness' in the water has been removed. It is calcium and magnesium ions that make the water hard. These react with soap to form a scum. If a soapless detergent is used instead, no scum forms and a lather forms immediately on agitation (see figure 6.7).

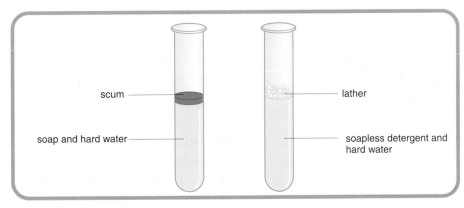

Figure 6.7 *Comparing soaps and soapless detergents*

Dry cleaning

What happens to clothes that are sent for 'dry cleaning' (see figure 6.8)? It may surprise you to know that *liquids* are used – so the process is not really dry! The liquids used in **dry cleaning** are special. They must be very good at removing stains caused by oil and grease. Dry cleaning liquids are in fact *solvents* because they *dissolve* the oil and grease.

Figure 6.8 *Dry cleaning takes place in this machine*

Because dry cleaning machines are found in shops where people work, the solvents used must be safe. Here is a list of some of the properties of a good dry cleaning solvent:

- not very poisonous,
- does not catch fire easily,
- does not harm fabrics,
- vaporise easily.

It is important that a dry cleaning solvent should vaporise easily so that the clothes *dry* quickly.

Section Questions

1 Why is greasy hair difficult to clean with water alone?

2 Complete the following sentence:

A cleaning chemical, such as a shampoo, can break up grease into tiny ...., which can then ...... with water.

3 Why are soaps less efficient than soapless detergents for cleaning skin and clothes using hard water?

4 Complete the following sentence:

The solvents used in dry cleaning are particularly good at dissolving ...... and ....

Factors that affect lathering

Information

The aim of the experiment is to investigate a factor that might affect
the amount of lather produced when detergents are shaken with water.

The factors that can be investigated are as follows:

- type of detergent,
- volume of detergent,
- temperature of the water,
- amount of shaking,
- volume of water.

To make an investigation fair, only one factor must
be changed during the experiments. All of the others
must stay the same. In this investigation, the factor
that you will be changing is the 'type of detergent'.

What to do

1 Add water to a small beaker so that it is half full.

2 Use a syringe to measure 3 cm³ of the water into a
test tube (see figure 6.9).

3 Add two drops of automatic washing detergent
solution to the test tube (see figure 6.10).

4 Put a stopper in the test tube and then shake it hard
for 15 seconds (see figure 6.11).

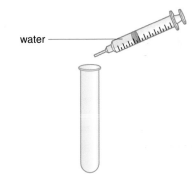

Figure 6.9 *Adding water to the test tube*

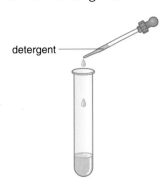

Figure 6.10 *Adding detergent to the test tube*

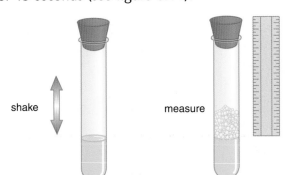

Figure 6.11 *Shaking
the test tube*

Figure 6.12 *Measuring
the height of the lather*

5 Allow the mixture to settle for 15 seconds and then measure
the height of the lather using a ruler (see figure 6.12).

6 Record your result.

7 Rinse out the test tube.

8 Repeat the process with the same detergent, starting at step 2.

9 Carry out the procedure twice more, using non-automatic washing detergent solution. Record your results.

Results table

type of detergent	height of lather / cm		
	result 1	result 2	average
automatic	1.0	1.4	1.2
non-automatic	5.3	5.7	5.5

Conclusion

The type of detergent used affects the amount of lather produced.

Clothing

Figure 6.13 *Wool comes mainly from sheep*

Most clothing is made from cloth or **fabric**. Clothing fabrics are made from thin strands of material called **fibres**. There are two main types of fibres.

Natural fibres come from plants and animals. Wool fibres, for example, come mainly from sheep (see figures 6.13 and 6.14). Cotton fibres are also natural and come from the cotton plant (see figure 6.15). Silk is produced by the silk worm (see figure 6.16).

Figure 6.14 *Alpacas produce very fine wool*

Figure 6.15 *Cotton comes from the cotton plant*

Figure 6.16 *Silk is produced by the silk worm*

Synthetic fibres are ones that are made by the chemical industry. Nylon and polyesters are examples of synthetic fibres. Nylon is often referred to as 'polyamide' on clothing labels. Terylene is the most common polyester fibre that is used for clothing. Nylon and polyester are often used together, for example in anoraks. We wear clothes made of a variety of different fibres (see figures 6.17 and 6.18).

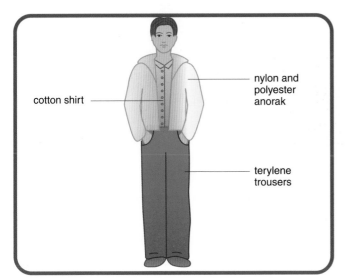

Figure 6.17 *Boys wear a variety of fibres …*

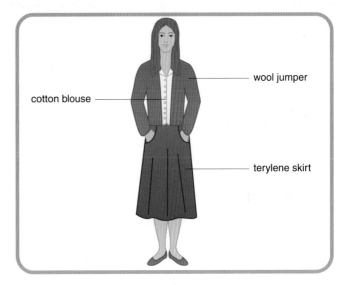

Figure 6.18 *… and so do girls*

Properties of different fibres

Which do you think is warmer, a cotton jumper or a wool jumper? Wool keeps us warmer than cotton. This is an important property of wool. Cotton clothes are better for keeping us cool. Unfortunately, both wool and cotton can make us sweaty when we are hot. Terylene is much better than either cotton or wool for sports clothes (see figure 6.19). This is because it lets water molecules pass through it easily.

Figure 6.19 *Terylene sportswear helps to keep athletes as cool as possible*

Synthetic fibres can be used to produce fibres with specific properties. A special kind of nylon, for example, is used to make toothbrush bristles (see figure 6.20).

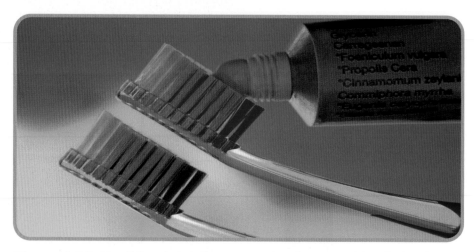

Figure 6.20 *Toothbrush bristles are made of a special kind of nylon*

The polyester fibre Terylene is particularly versatile. It can be used for many purposes. Gloves made of Terylene are good at keeping hands warm (see figure 6.21). Vests, as worn by hill walkers in winter, allow perspiration to pass through them (see figure 6.22). This means that they do not become wet, as a cotton vest would. T-shirts have traditionally been made of cotton. Terylene may replace cotton for this purpose for the same reason.

Figure 6.21 *Gloves made of Terylene are good at keeping hands warm*

Figure 6.22 *This Terylene vest lets perspiration pass through it easily*

Polypropene can be used to make hard-wearing, stain-resistant carpets. It can also be used as synthetic grass for hockey pitches (see figure 6.23).

Figure 6.23 *Hockey being played on the synthetic fibre polypropene*

Section Questions

5 Cotton, polyester, wool, nylon and silk are fibres. Give this information as a table using the two headings 'natural fibres' and 'synthetic fibres'.

6 What is meant by the term 'synthetic fibre'?

Fibres are made up of polymers

All fibres are made up of long chain molecules. These long chain molecules are called **polymers** (see figure 6.24).

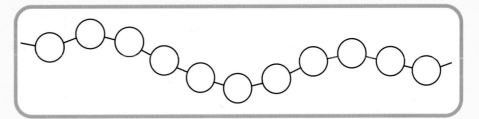

Figure 6.24 *Part of a polymer molecule in a fibre*

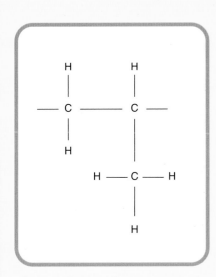

Figure 6.25 *The repeating unit in polypropene*

In a polymer, a basic group of atoms forms a repeating unit. This is shown in figure 6.24 as $\cdot$O$\cdot$. In the case of polypropene, this repeating unit is shown in figure 6.25.

103

It is the forces of attraction between polymer molecules that give the fibres their strength. The forces of attraction between nylon and Terylene polymers are stronger than those between wool and cotton polymers. Nylon and Terylene fibres are therefore stronger than wool and cotton fibres.

Which fibres absorb water?

If some water is spilled, what fibre is best at absorbing it? Simple experiments show that cotton is best. Wool is quite good, but nylon and Terylene are not very absorbent. The cotton and wool fibres form strong bonds with water molecules. They are not only absorbent but also do not dry quickly either (see figure 6.26).

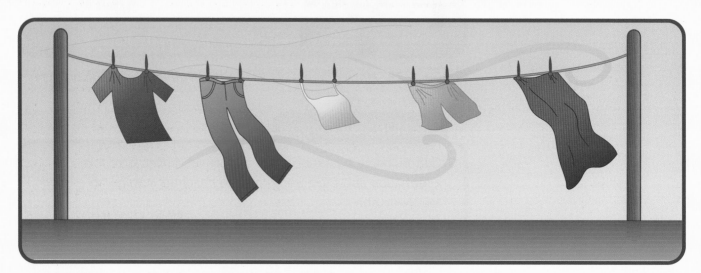

Figure 6.26 *Some fibres dry quicker than others*

Clothes made of cotton are slow to dry. Think how long it takes to dry denim jeans. Denim is made of cotton. Wool also takes quite a long time to dry. Nylon and Terylene dry quickly and easily drip-dry.

Wool and cotton can feel 'sweaty' next to the skin because they soak up perspiration. Terylene is best next to the skin because it lets water molecules pass through it easily.

Section Questions

7 Some properties of four fibres are shown in the table. The more stars, the better the fibre.

	wool	cotton	nylon	Terylene
strength	**	***	****	****
absorbency	***	****	**	*
drying ease	**	*	****	****

a) Which two fibres are strong and dry easily when wet?
b) Which fibre would be best for making a dish towel, used to dry dishes?
c) Which two fibres form strong bonds with water molecules?

Treating fabrics

Cotton fabric catches fire easily. Some children have been badly burned because of wearing cotton clothes too near to an open fire. Fortunately, cotton can be made **flame-resistant** by treating it with special chemicals. One such chemical is sodium silicate solution. You will find symbols, like the one shown in figure 6.27, to show that mattresses, for example, have been treated to make them flame-resistant.

On some clothes you may see labels that a fabric protector called Teflon has been used on them. On the reverse of the label shown in figure 6.28, the following claim is made: 'Water rolls off – soils and stains just wipe away'. Teflon evidently makes the fabric resistant to water, stains and dirt.

Chemicals called **silicones** are very good for water-proofing fabrics. Anoraks are often sold with a water-repellent coating. However, washing can gradually remove this, and the anorak must be sprayed with silicone to make it water-repellent again (see figure 6.29). The same material can also be used on footwear to make shoes and boots water-proof.

Figure 6.27 *This label shows that the fabric used is flame-resistant*

Figure 6.28 *Have you seen a label like this on clothing?*

Figure 6.29 *This silicone spray can make an anorak water-proof*

Dyeing fibres

Most fibres, such as cotton, nylon, polyester and silk, are white. Wool is usually white, but can have natural colour. Coloured fibres can be produced using **dyes**. Dyes are coloured materials that can be used to give colour to a fibre. They can be natural or synthetic. Some sources of natural plant dyes are shown in figures 6.30 and 6.31.

Figure 6.30 *Elderberries and blackberries can be used as dyes …*

Figure 6.31 *… as can beetroot and onions*

In order to extract a dye from plant material, it is chopped up, covered in water and boiled gently for about thirty minutes (see figure 6.32). The solid material (for example, onion skins) is then filtered off (see figure 6.33), leaving the dye in solution.

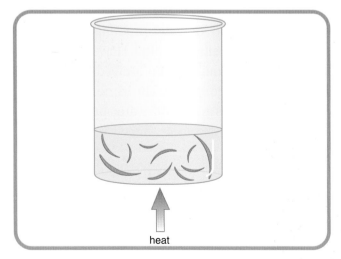

Figure 6.32 *Heating the chopped up plant material*

Figure 6.33 *Filtering off the solid material*

The undyed fibres are placed in the dye solution and boiled gently for about ten minutes (see figure 6.34). Longer boiling produces more intense colours. The fibres are removed, rinsed in cold water and allowed to dry. Wool that has been dyed in this way is shown in figure 6.35.

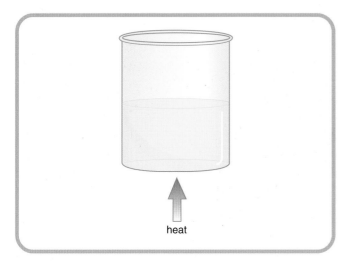

Figure 6.34 *Dyeing the fibres*

Figure 6.35 *This wool has been dyed using onion skins*

Making the dye more permanent

Dyeing fibres and fabrics with natural dyes can result in the colour being easily washed out. Natural dyes can be fixed more permanently onto fibres by treating the fibre with a **mordant** before dyeing. Alum (aluminium potassium sulphate) is a suitable mordant for wool. Before dyeing, the wool is heated with an alum solution.

Synthetic dyes

Perhaps you have used a synthetic dye to restore the original blue colour to faded jeans. A suitable dye is shown in figure 6.36. Synthetic dyes tend to bond strongly onto fibres and do not need to be treated with a mordant.

Figure 6.36 *This blue dye can restore the colour to faded jeans*

Identifying fibres

Shirlastain A is a mixture of three synthetic dyes. A simple chromatography experiment can show their colours (see figure 6.37).

Figure 6.37 *Shirlastain A contains three dyes*

The dyes in Shirlastain A are very selective. Those fibres that are chemically different respond differently to the dyes. Figure 6.38 shows the result of dyeing a mixture of fibres with Shirlastain A.

Using Shirlastain A, the results shown in table 6.1 are obtained.

Figure 6.38 *Different dyes adhere to different fibres*

fibre	colour
wool	brown
cotton	blue
Terylene	almost no colour

Table 6.1 *Results of dyeing with Shirlastain A*

Fuels

Fire

For a fire to burn, three things are needed.
- Fuel – a substance that burns to produce energy.
- Oxygen – when a substance burns, it reacts with oxygen.
- Heat – this is needed to start a fire and to keep it burning.

North Sea gas is the fuel for most of the bunsen burners in school laboratories (see figure 7.1).

We can show that oxygen is used up when a candle burns by the experiment shown in figure 7.2.

Figure 7.1 *A bunsen flame*

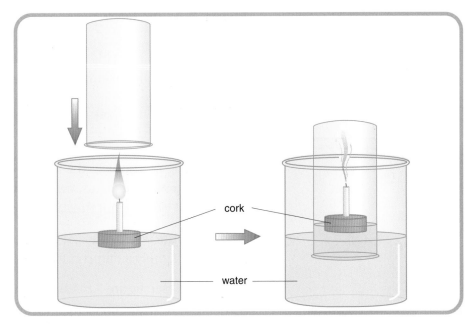

Figure 7.2 *Which gas makes up one-fifth of the air?*

Oxygen makes up one-fifth of the air. Therefore, oxygen must be the gas that is used up during burning.

Another word for burning is **combustion**.

The fire triangle

The three things that are needed for burning, a **fuel**, **oxygen** and **heat**, make up the three sides of the **fire triangle** (see figure 7.3).

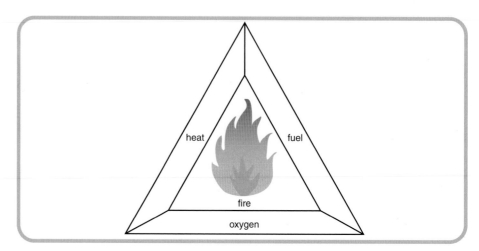

Figure 7.3 *The fire triangle*

If you take away one side of the triangle, it will collapse. This means that if you deprive a fire of either fuel or oxygen or heat, then it will go out.

Fighting fires

Fire-fighting methods in the laboratory and in the home include the use of a fire blanket, sand, water, carbon dioxide and foam. Different methods of putting out fires are used in different situations.

A chip pan fire (see figure 7.4) can be put out by covering it with a fire blanket, or simply a damp cloth. This prevents oxygen getting to the fire.

Forest fires (see figure 7.5) are fought with water, sometimes dropped from aircraft. Water takes heat away from a fire. There may also be time to chop down some of the trees. This deprives the fire of fuel.

Figure 7.4 *A chip pan fire*

Figure 7.5 *A forest fire*

Throwing sand over burning petrol (see figure 7.6) will put out the fire. This will keep out the oxygen and take away heat.

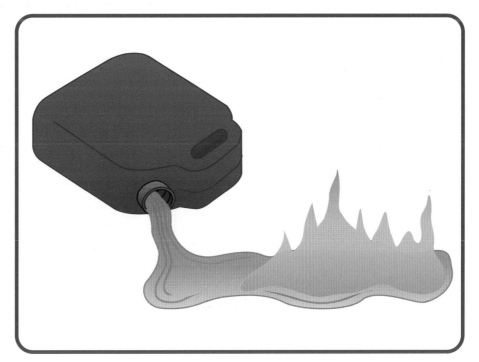

Figure 7.6 *A petrol fire*

When an aircraft catches fire at an airport (see figure 7.7), foam is used to put out the fire. The foam smothers the flames, preventing oxygen from reaching the burning fuel.

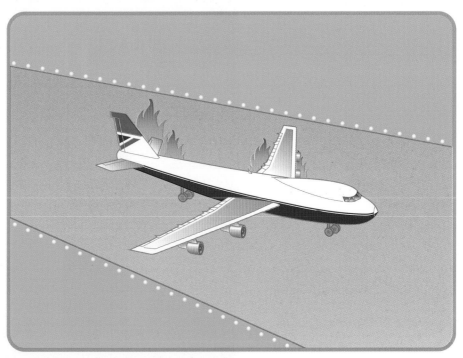

Figure 7.7 *Fortunately aircraft rarely catch fire*

Pyjamas and nightdresses are often made of cotton. Unfortunately, cotton catches fire easily (see figure 7.8). If a fire blanket is not available, the person can be wrapped in a rug. This will stop oxygen getting to the fire.

Figure 7.8 *Cotton pyjamas should be kept away from flames*

Clothes shops often have carbon dioxide fire extinguishers ready for use (see figure 7.9). Carbon dioxide is a dense gas that stops oxygen getting to the fire. Water can also be used, but is more messy.

Figure 7.9 *Carbon dioxide gas is found in many fire extinguishers*

What not to use

Did you know that water must *not* be used with oil, petrol or electrical fires? Oil and petrol float on water, so using water just spreads the fire.

Water should not be used on electrical fires, because water conducts electricity.

Section Questions

1 What is removed from the fire triangle when water is used to put out a fire made of burning wood?

2 Why should you *not* throw water on a chip pan fire?

Finite resources

Fossil fuels

Coal, oil, peat and natural gas are called **fossil fuels**. This is because they were formed from plant and animal material over a very long time.

Coal was formed from trees and other vegetation growing in swampy areas (see figure 7.10). When these died, they fell into the swamp and became covered by layers of mud. Gradually these plant remains turned into coal.

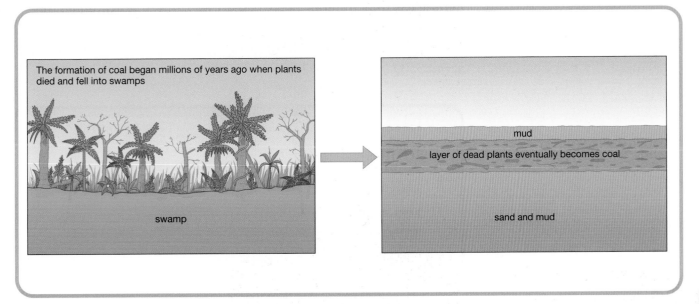

Figure 7.10 *Coal was formed from plant material like this*

Oil and **natural gas** were formed from tiny sea creatures and plants (see figure 7.11). When these died, they sank to the sea bed. Eventually they were buried under layers of sand and other materials. Over a long period of time, they turned into oil and gas.

Figure 7.11 *Oil and natural gas were formed from tiny sea creatures and plants like these*

Figure 7.12 *Many crofters traditionally cut peat for use as a fuel*

Peat was formed from plant material growing in swampy areas. Unlike coal and oil, peat has not been buried underground (see figure 7.12). Peat was formed more recently than the other fossil fuels.

A fuel crisis

We are using up our fossil fuels at an alarming rate. Coal, oil and natural gas took millions of years to form. Sadly, it is estimated that oil may have run out by about 2050. Natural gas may only last until about 2040. The most plentiful fossil fuel is coal. This may last until about 2300 (see figure 7.13).

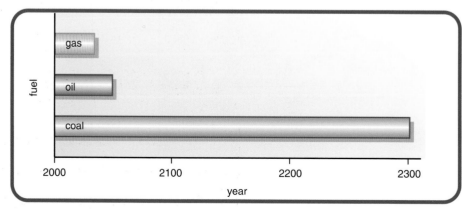

Figure 7.13 *How long will fossil fuels last?*

The fossil fuels are **finite resources**. This means that there is only a limited supply of them. Once used up, they cannot be replaced. There may be a fuel crisis soon because, when the fossil fuels have been used up, alternative sources of energy must be found.

Problems of transportation

The fossil fuels are not always found where they are needed. As a result, they must be transported, often by ship. If a ship carrying coal sinks, the coal sinks also and causes no great problem. Unfortunately, some of the largest ships carry oil. If one of these oil tankers sinks, the oil will float on the sea, causing a great deal of damage to marine life and to the environment (see figure 7.14). At sea, birds are particularly at risk from oil slicks. When the oil reaches the coast, beaches are ruined and tourism suffers.

Figure 7.14 *Oil spillages cause a great deal of harm*

Section Questions

3 a) Name four fossil fuels.
   b) What were fossil fuels formed from?

4 Fossil fuels are finite resources. What does 'finite' mean?

5 Why could our over-use of fossil fuels lead to a fuel crisis?

Figure 7.15 *The hydrocarbon butane provides the heat for this barbecue*

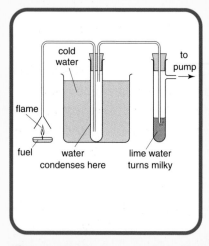

Figure 7.16 *Investigating the products of burning a hydrocarbon*

Hydrocarbons

The compounds that are found in fossil fuels are mainly **hydrocarbons**. A hydrocarbon is a compound that contains hydrogen and carbon only. Examples of hydrocarbons include methane, CH_4, and butane, C_4H_{10} (see figure 7.15). Both are found in natural gas.

Experiments, such as the one shown in figure 7.16, show that, when hydrocarbon fuels burn in a plentiful supply of air, carbon dioxide and water are produced.

The gases from the burning hydrocarbon are drawn through the apparatus using a pump. Carbon dioxide in the gas mixture turns the lime water milky. This is the chemical test for carbon dioxide. Condensation appears in the cooled test tube as water vapour from the burning hydrocarbon comes into contact with the cold glass. If enough is collected, this colourless liquid can be proved to be water by measuring its boiling point and freezing point. Water boils at 100°C and freezes at 0°C.

The word equation for the burning of a hydrocarbon in a plentiful supply of air is as follows:

hydrocarbon + oxygen → carbon dioxide + water

For example:

butane + oxygen → carbon dioxide + water

Section Questions

6 a) Which elements are present in hydrocarbons?
   b) What products are formed when a hydrocarbon burns in plenty of air?

Renewable resources

As fossil fuels run out, new fuels must be found to replace them. If possible, these should be **renewable resources**. This means that they can be replaced. Methane, ethanol and hydrogen are renewable sources of energy.

Methane is found in **biogas**, which can be made by the decomposition of plant material. The plant material, such as grass and vegetable waste (see figure 7.16), is allowed to decompose in the absence of air. The biogas so produced can contain up to 75% of methane. Methane is used as a fuel for homes and factories (see figure 7.17). It is also used in some power stations to generate electricity. Animal manure can also be used to produce biogas.

Figure 7.17 *Plant waste like this …*

Figure 7.18 *… can be turned into methane gas*

Ethanol or 'alcohol' is obtained from sugar cane by fermentation (see figure 7.19). Mixed with yeast and dissolved in water, sugar reacts to produce ethanol. This can be removed from the other substances by distillation. The process is similar to that used in making whisky. The ethanol produced can be mixed with petrol to make a fuel for cars (see figure 7.20).

Figure 7.19 *Sugar can be fermented …*

Figure 7.20 *… to make ethanol – a fuel for cars*

Water, H_2O, is a very common compound. It covers about four-fifths of the Earth's surface (see figure 7.21). **Hydrogen**, which can be obtained from water using electricity, is a likely fuel for the future. Already some cars are running on hydrogen (see figure 7.22).

Figure 7.21 *The Earth has a lot of water …*

Figure 7.22 *… which can provide hydrogen for cars to run on*

When hydrogen is used as a fuel in cars, the only product is water. The reaction taking place is as follows:

hydrogen + oxygen → water

Section Questions

7 Methane is present in biogas.
a) What is biogas?
b) Why is methane said to be a renewable resource?
c) Name two other renewable resources.

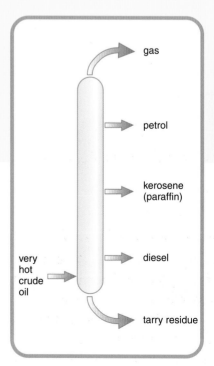

Figure 7.23 *A fractionating tower*

Processing crude oil

Crude oil is the most important of the fossil fuels. We use more of it than of all the others put together. It is a mixture of thousands of compounds, mainly hydrocarbons. But it is almost useless until it has been put through various processes, which we call **refining**.

The first stage in the refining of crude oil is **fractional distillation**. This is a process of distillation based on differences in boiling points. The crude oil is separated into **fractions**. A fraction is a group of hydrocarbon molecules that have boiling points within a certain range. A simplified version of a fractionating tower is shown in figure 7.23.

The different fractions, gas, petrol, kerosene and diesel, are used as fuels. The gas fraction consists mainly of propane and butane. Propane is sold in red cylinders, while butane is put into blue ones (see figure 7.24).

Propane and butane are used mainly for heating and cooking. Petrol, kerosene and diesel provide energy for transporting people. These uses are shown in table 7.1.

Figure 7.24 *Propane and butane*

fuel	use
propane	heating and cooking
butane	heating and cooking
petrol	motor vehicles
kerosene	jet aircraft
diesel	motor vehicles

Table 7.1 *Uses of some fuels*

A mixture of propane and butane is also sold as a fuel for motor vehicles. You may have seen this at some filling stations (see figure 7.25).

Figure 7.25 *A mixture of propane and butane is for sale from this pump*

Section Questions

8 Copy and complete the following using words from the word bank.

fuels	jet	hydrocarbons
diesel	boiling	petrol
		distillation

Crude oil is a mixture of .... It can be separated into fractions using fractional .... The fractions have different .... point ranges. The main use of the fractions is as .... These include ...., ... and kerosene. Kerosene is used in ...... aircraft.

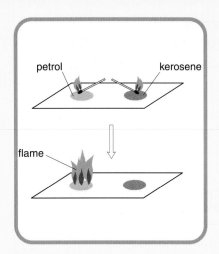

Figure 7.26 *Petrol is more flammable than kerosene*

Properties of the fractions

Boiling range and **viscosity** (how 'thick' a fraction is) increase with molecular size. This is because the bonds *between* molecules increase in strength as the molecules become bigger. More energy and therefore a higher temperature is needed to turn fractions with larger molecules from liquid to gas. Also, as the bonds between molecules become stronger, a liquid fraction will become less 'runny' and more viscous.

Ease of vaporisation and **flammability** are also linked to molecular size. It is the vapour or gas from a hydrocarbon that catches fire. Kerosene is made of larger molecules than petrol. As a result, petrol vaporises more easily and is more flammable. Figure 7.26 shows what happens when lit tapers are held above kerosene and petrol.

Summary

gas	**petrol**	**kerosene**	**diesel**

⟶ increasing molecular size ⟶

⟶ increasing boiling range ⟶

⟶ increasing viscosity ⟶

⟶ decreasing flammability ⟶

⟶ decreasing ease of vaporisation ⟶

Properties and uses

The uses of the fractions are related to their properties. For example, petrol, kerosene and diesel have low enough viscosity to flow through pipes into the combustion chambers of engines. Once there, they are flammable enough to ignite. Lubricating oil and bitumen, which are obtained from the tarry residue, are much more viscous and much less flammable. Lubricating oil therefore clings to the moving parts of engines. Bitumen does not become runny on road surfaces (except during *very* hot weather).

Supply and demand

Fractional distillation of crude oil produces more long chain hydrocarbons than are useful for present-day industrial purposes. The supply and demand for fuel oil and petrol are compared in figure 7.27. (Fuel oil is obtained from the tarry residue and is used in ship's engines.)

Figure 7.27 *Comparing the supply and demand of petrol and fuel oil*

The unwanted large hydrocarbon molecules in fuel oil can be broken down into smaller more useful molecules using a process called **cracking**. The mixture of smaller molecules is fractionally distilled. This produces, among other fractions, more petrol. The main purpose of the cracking of fuel oil in a refinery is to produce more petrol and diesel.

A cracking experiment

A lubricating oil, like engine oil or medicinal paraffin, can be cracked using the apparatus shown in figure 7.28. Some of the smaller hydrocarbons produced are gases. The 'cracked gas' burns when a lit taper is applied to it.

The aluminium oxide used in the experiment is present as a catalyst.

When cracked, a large hydrocarbon molecule can break up in many ways. One example is shown below:

$$C_{25}H_{52} \longrightarrow C_{16}H_{34} + C_6H_{14} + C_3H_6$$

found in fuel oil   found in diesel   found in petrol   a gas

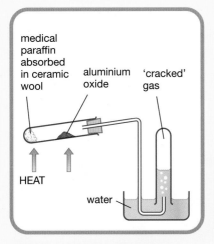

Figure 7.28 *A laboratory cracking experiment*

medical paraffin absorbed in ceramic wool

aluminium oxide

'cracked' gas

HEAT

water

Section Questions

9 Lubricating oil is more viscous than kerosene. What does this mean?

10 Petrol vaporises and burns more easily than diesel. Which is made of the bigger molecules?

11 a) In terms of supply and demand, why is fuel oil cracked to make petrol and diesel?
   b) What is cracking?

12 In the following cracking reaction, a hydrocarbon molecule is broken up into two smaller ones:

$$C_6H_{14} \rightarrow C_4H_{10} + X$$

Give the chemical formula for X.

Fuels and pollution

A **pollutant** is a substance that harms the environment. The burning of hydrocarbon fuels can produce many pollutants.

Carbon monoxide

Any hydrocarbon fuel can produce poisonous carbon monoxide if it burns in a low supply of oxygen. Gas fires, cookers and boilers all need a good supply of oxygen. All petrol and diesel engines produce carbon monoxide. They should not be run in a confined space such as a garage (see figure 7.29).

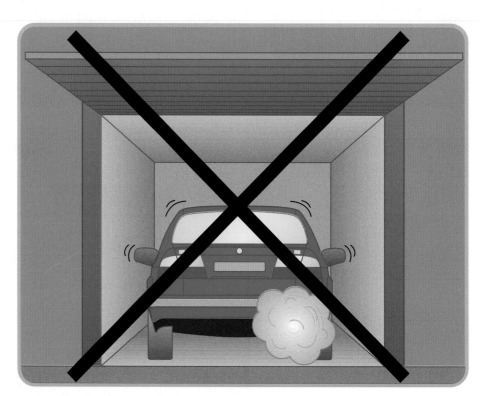

Figure 7.29 *Never run a car engine in a confined space*

Carbon particles

Carbon particles, known as 'soot', can also be produced when a hydrocarbon fuel burns incompletely. Candles contain large hydrocarbon molecules. If a white tile is put into the yellow part of the flame, it soon becomes coated with black soot (see figure 7.30). This shows that carbon is being produced in that part of the flame.

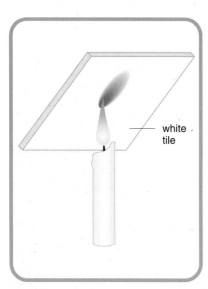

white tile

Figure 7.30 *The yellow part of a candle flame contains soot*

The soot particles that are produced by the incomplete burning of diesel are known to be harmful. Studies of taxi drivers working in towns and cities have shown increased ill health due to these particles (see figure 7.31).

Figure 7.31 *Soot pollution from diesel engines can make taxi drivers ill*

Sulphur dioxide

Coal, crude oil and natural gas all contain sulphur compounds. When these burn, they produce sulphur dioxide, which is poisonous, acidic and irritating. Most of the sulphur is removed from petrol and diesel, but not from coal. Coal-fired power stations (see figure 7.32) produce a lot of sulphur dioxide, but not all remove it from their flu gases.

Figure 7.32 *Over five million tonnes of coal are burned each year at Longannet power station*

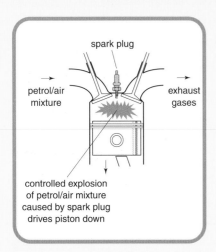

Figure 7.33 *Inside a petrol engine*

Nitrogen dioxide

In petrol engines, and to a lesser extent in diesel engines, nitrogen and oxygen react to produce nitrogen dioxide. This gas is also poisonous, acidic and irritating. In a petrol engine (see figure 7.33), spark plugs provide electrical sparks to ignite a mixture of petrol vapour and air. Near the spark, the temperature is so high that nitrogen and oxygen react as well.

Diesel engines do not have spark plugs and do not reach such a high temperature as petrol engines. They therefore cause less reaction between nitrogen and oxygen molecules.

Benzene

Some years ago lead compounds were added to petrol. This improved the way in which it burned in engines. Unfortunately, these lead compounds caused pollution. A decision was taken to stop adding them to petrol.

In place of the lead compounds, the petrol was further refined and called 'unleaded' petrol. This involved changing the structure of the hydrocarbon molecules. One of the molecules produced is the very toxic (poisonous) compound benzene. The oil companies now remove as much of this as possible at the refinery (see figure 7.34).

Figure 7.34 *In this part of the BP refinery at Grangemouth, benzene is removed from petrol*

Catalytic converters

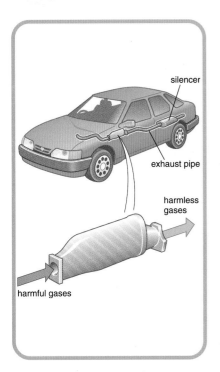

Figure 7.35 *A catalytic converter*

All new cars are now fitted with catalytic converters in their exhaust systems (see figure 7.35). These, when hot, help to reduce pollution from car engines. They contain some expensive transition metals as catalysts. These cause harmful gases, like carbon monoxide, to be converted into less harmful ones. Carbon monoxide is converted into carbon dioxide, which is not poisonous.

Section Questions

13 What causes hydrocarbon fuels to produce carbon and carbon monoxide when they burn?

14 What harmful *solid* particles can be produced by the incomplete combustion of diesel?

15 a) Which acidic pollutant gas is produced by the burning of compounds that contain sulphur?
b) Which two elements join at high temperatures in car engines to form nitrogen dioxide?

16 Why is as much benzene as possible removed from petrol?

ACCESS 3 Subsection Test: Personal needs and Fuels

Part A

This part of the paper consists of four questions and is worth 4 marks.

1 Which word describes the behaviour of oil and grease when they are added to water? (1)

soluble or **insoluble**

2 Which of the following words best describes a silk fibre? (1)

natural or **synthetic**

3 Which gas do burning substances react with? (1)

nitrogen or **oxygen**

4 Which of the following is an example of a fossil fuel? (1)

wood or **peat**

Part B

This part of the paper is worth 6 marks.

5 Cleaning chemicals break up oil and grease into small droplets. Name a manufactured product that contains cleaning chemicals. (1)

6 The results of testing some fibres are shown in the table.

fibre	heat resistance	drying rate
wool	poor	poor
nylon	poor	good
cotton	good	poor
Terylene	moderate	good

Which fibre could withstand the use of a hot iron, but would dry only slowly when wet? (1)

7 Putting a damp towel over a chip pan fire can successfully put out the flames. Explain how this technique works. (1)

8 The index of a book contains the following references to fuels.

subject	page number
coal	16
gas	5
oil	72
peat	38
wood	29

On which page would you expect to find information about natural gas? (1)

9 The world's fossil fuels are finite resources, but it is expected that coal will outlast oil and natural gas. The graph below gives information about oil and natural gas.

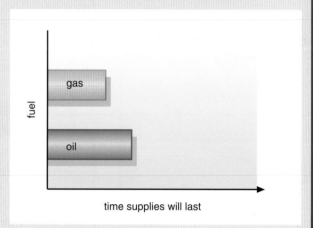

Copy the graph and add to it a bar to show how long supplies of coal are likely to last. (1)

10 Air pollution from cars has been reduced by the use of catalytic converters.

What happens to pollutant gases when they pass through a catalytic converter? (1)

Total 10 marks

8 Plastics

Although most crude oil is turned into fuels, some is made into other things such as **plastics**. Plastics are synthetic materials made by the chemical industry, using crude oil as the main raw material.

Uses of plastics

So many of the things we use are made of plastics that it is difficult to imagine life without them. Roller-blades, helmets, mobile phones, bottles, kettles, DVDs (and their boxes), cameras, watches, and many more items, are all made of various plastics (see figures 8.1 to 8.4).

Figure 8.1 *Many parts of these roller-blades are made of plastics*

Figure 8.3 *Helmets are made of the very strong plastic Kevlar*

Figure 8.4 *The protective gear worn by American football players is made of plastic*

Plastics are used for many different jobs because they have many useful **properties**. The properties of materials describe what they are like and how they behave. The everyday uses of plastics are related to their properties. This is shown in table 8.1.

plastic	property	use
polythene	flexible	detergent bottles
Kevlar	very strong	helmets
Formica	heat-proof	kitchen work-tops
polystyrene	heat insulator	fast-food packaging

Table 8.1 *Properties and uses of some plastics*

Figure 8.2 *This is a 'PET' bottle – it is made of polyester*

Perspex is a plastic that can be used in place of glass. It is stronger than glass and does not shatter (see figure 8.5).

Figure 8.5 *This bus shelter has Perspex glazing*

Polyvinyl chloride, better known as PVC, has many uses. The outside doors, window frames and gutter pipes of the house in figure 8.6 are all made of PVC.

Figure 8.6 *A lot of PVC is used in building modern houses*

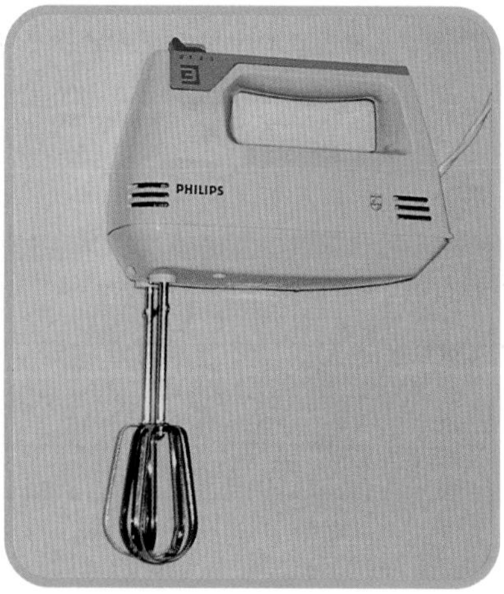

Figure 8.7 *This food mixer has nylon gear wheels*

Nylon is another strong plastic. Curtain hooks and the gear wheels of food mixers are two of its uses (see figure 8.7).

Bakelite was one of the first plastics to be made. It is very heat-resistant. You may still have some bakelite electrical fittings in your home, if it is old enough.

Most plastics become hard and brittle at low temperatures, but not silicones. Tubing made of silicones stays flexible even when it is very cold.

Section Questions

1 Make a list of four useful properties of the plastic used to make shopping bags, like the one shown.

Plastics versus traditional materials

For some uses, plastics have advantages over traditional materials and vice versa.

PVC is used in place of wood for window frames and doors. This is because PVC does not rot and does not require painting. The same plastic is used for gutter pipes, replacing iron. This is because it does not rust. However, PVC is not as strong as iron. A ladder, for example, is more likely to damage a plastic gutter than an iron one.

Plastic bottles, made of polyester (PET), are preferred to ones made of glass. This is because they are lighter and shatterproof.

Maps printed on polythene are waterproof. Hill-walkers tend to prefer them to the traditional maps, which are printed on paper (see figure 8.8).

Polythene pipes are now used to carry mains water (coloured blue) and mains gas (coloured yellow). Unlike the iron pipes they have replaced, they do not rust.

Figure 8.8 *Which map is printed on plastic?*

Figure 8.9 *Would a door made of PVC look better than this one made of wood?*

Figure 8.10 *Plastic waste can cause environmental problems*

Sometimes a traditional or natural material is preferred to one made of plastic. Natural flowers are usually preferred to plastic ones. Some people think that the natural grain of an outside door made of wood looks better than the smooth finish of one made of PVC (see figure 8.9).

Section Questions

2 Give three reasons, based on differences in properties, why PVC tends to be used in place of wood for window frames in houses.

Plastics and pollution

Most plastics are **durable**, which means that they last for a long time. This is often a useful property, but it can also cause problems. Plastic litter can remain in the environment for a long time. In contrast, natural materials like wood and paper are **biodegradable**. This means that they are broken down by bacteria in the soil and rot away. The lightweight nature of plastics can also be a problem. Crisp packets and other plastic waste, such as polythene sheets, can be blown about and left caught in trees, bushes and fences (see figure 8.10).

Some biodegradable plastics have been developed by chemists. One of these is called **Biopol**. Because they rot away, they cause fewer problems in the environment. As yet, biodegradable plastics are not very common. This is because plastics are too valuable to be thrown away. They should be recycled instead.

The burning of plastics can produce a lot of pollution. They all contain carbon and, if insufficient oxygen is present, they produce poisonous carbon monoxide when they burn. Different plastics cause a variety of problems when they burn. Polystyrene, for example, produces thick black sooty smoke when it burns. PVC contains chlorine and produces poisonous hydrogen chloride gas. Some plastics can produce the even more dangerous gas, hydrogen cyanide.

Section Questions

3 Most plastics are not biodegradable. What does this mean?

4 Why does the lightweight nature of plastics cause problems in the environment?

5 Name the poisonous gas that all plastics can produce if they burn with not enough oxygen present.

Disposing of plastic waste

The options for disposing of our plastic waste include incineration (burning), recycling and burying.

Incineration could provide a useful supply of heat energy. Unfortunately, poisonous gases would be released.

Recycling is probably the best option. This would help to extend the limited supplies of crude oil. The problem with this method of disposal is that many different plastics are in common use.

Burying plastic is wasteful. Also, landfill sites would fill up quicker and most plastics are not biodegradable.

More about recycling

You will probably hear more about recycling in the years to come. Most European countries recycle much more of their waste materials than we do. However, some companies are already involved in recycling. The supermarkets, for example, collect used polythene shopping bags (see figure 8.11). This plastic can be recycled as bin-liners. Some fast-food chains collect their polystyrene waste. Recycled polystyrene can reappear as plant pots.

Planning for the future

Chemists are already looking ahead to when the supplies of crude oil dwindle. Renewable sources of plastics are already being examined. These include **starch**, which, because plants produce it, should never run out.

Figure 8.11 *Many supermarkets provide collecting points for polythene shopping bags*

Section Questions

6 Why should plastics be recycled rather than incinerated or buried?

7 Why are chemists already looking for renewable sources of plastics?

Thermoplastics and thermosetting plastics

Some plastics soften and melt when heated, while others do not. Plastics are classified based on this property.

■ **Thermoplastics** soften on heating and can be reshaped.
■ **Thermosetting plastics** do not soften on heating and cannot be reshaped.

Examples of thermoplastics include polythene, polystyrene, Perspex, PVC and nylon. Because they soften on heating, and can be reshaped, all of them can be recycled. With care, you can show that the polystyrene case of a ball-point pen will soften in very hot water. It can then be remoulded into a different shape. Figure 8.12 shows how this can be done.

Figure 8.12 *Ball-point pen cases are made of polystyrene, which is a thermoplastic*

A simple test to distinguish between a thermoplastic and a thermosetting plastic is to touch it with a hot iron nail. If the plastic softens and melts, then it is a thermoplastic.

Thermosetting plastics can withstand much higher temperatures than thermoplastics. They are used where heat resistance is important. Formica, for example, is used for kitchen work surfaces.

Dark brown Bakelite was one of the first thermosetting plastics to be made. It was widely used for light fittings. Urea-methanal has largely taken the place of Bakelite because its normally white appearance is preferable to the dark brown of Bakelite (see figure 8.13). Both Bakelite and urea-methanal are excellent heat and electrical insulators. Along with their heat-resistant properties, this makes them ideal for use as electrical fittings.

Figure 8.13 *The plug and socket are made of thermosetting urea-methanal plastic*

Unfortunately the fact that thermosetting plastics do not soften on heating makes them difficult to recycle.

Section Questions

8 How would you distinguish between a thermoplastic and a thermosetting plastic? Describe your test and the expected results.

9 Why is it all right to make an ice-cube tray out a thermoplastic, but a frying-pan handle must be made of a thermosetting plastic?

Making plastics

Plastics are made up of very large molecules called **polymers**. These are made by joining together many small molecules called **monomers**. The process of making a large polymer molecule by joining together many small monomer molecules is called **polymerisation**.

In one polymerisation reaction, many ethene monomers join to form the polymer poly(ethene). The reaction is shown below:

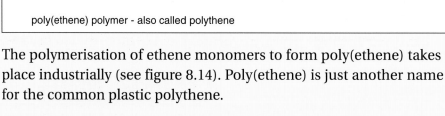

ethene monomers

poly(ethene) polymer - also called polythene

The polymerisation of ethene monomers to form poly(ethene) takes place industrially (see figure 8.14). Poly(ethene) is just another name for the common plastic polythene.

The way in which polystyrene is formed from many styrene monomers is similar to the polymerisation of ethene. The only difference is that the formulae are more complicated.

Figure 8.14 *Ethene polymerisation takes place in this plant at Grangemouth*

Section Questions

10 Plastics are polymers. What are polymers?

11 a) Name the polymer that is made from the monomer ethene.
   b) Name the monomer that is polymerised to form polystyrene.

ACCESS 3 Subsection Test: Plastics

Part A

This part of the paper consists of four questions and is worth 4 marks.

1   What are most plastics made from?   (1)

   natural gas or **crude oil**

2   What property of plastics can cause environmental problems?   (1)

   light (low density) or **heavy (high density)**

3   Which of the following is a natural material?   (1)

   Bakelite or **wood**

4   Thermosetting plastics are suitable for use as electrical plugs and sockets because they are   (1)

   conductors or **insulators**

Part B

This part of the paper is worth 6 marks.

5   Polythene is a thermoplastic. What does this mean?   (1)

6   PVC is now widely used in place of wood for window frames. Name **two** advantages that PVC has over wood.   (1)

7   Polystyrene is used for packaging and insulation. Polyester is used for fibres and bottles. Use this information to complete the following table.   (1)

plastic	uses
polystyrene	
	fibres and bottles

8   Poly(propene) is a plastic with various uses as shown by the table below.

use	% of production
fibre	35
car bumpers, etc.	25
film	20
packaging	10
other uses	10

Present this information as a bar graph.   (2)

9   What else could be done with plastic waste apart from incinerating it or burying it?   (1)

Total 10 marks

Intermediate 1 Unit Test: Everyday Chemistry

Part A

This part consists of twelve questions and is worth 12 marks.

In questions 1 to 6 choose the correct word to complete the sentences.

1 A nickel–cadmium battery **can/cannot** be recharged. (1)

2 Fibres that form strong bonds with water molecules are **hard/easy** to drip-dry. (1)

3 Fractional distillation of crude oil produces more **short/long** chain hydrocarbons than are useful for present-day industrial uses. (1)

4 Polymer molecules are made from many **large/small** molecules called monomers. (1)

5 Iron does not rust when attached to **less/more** reactive metals. (1)

6 An example of a renewable source of energy is **coal/ethanol**. (1)

Questions 7 to 12 are multiple choice questions. Choose the correct letter.

7 Which of the following metals reacts with dilute acid? (1)
A copper
B zinc
C silver
D gold

8 Which of the following is *not* an alloy? (1)
A solder
B steel
C iron
D brass

9 Which of the following correctly describes the solubility of cleaning chemicals in water and oil? (1)

	water	oil
A	insoluble	soluble
B	soluble	insoluble
C	insoluble	insoluble
D	soluble	soluble

10 An example of a non-metal element which conducts electricity is (1)
A sulphur
B phosphorus
C calcium
D carbon

11 Water must *not* be used to put out a fire caused by the burning of (1)
A wood
B oil
C coal
D paper

12 Which of the following is *not* regarded as a finite resource? (1)
A coal
B wood
C natural gas
D crude oil

Part B

This part consists of eight questions and is worth 18 marks.

13 Magnesium reacts slowly with water to produce a colourless gas.

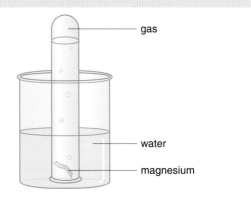

gas
water
magnesium

a) The colourless gas burned with a pop. Name this gas. (1)

b) Refer to page 6 of the data booklet and name a metal that would react faster with water than magnesium. (1)

14 70% of the polyester plastics produced are used for clothing, 10% is used for film and 20% for packaging.

a) Present this information as a labelled bar graph. (1)

b) An example of a use for polyesters as packaging is as bottles. Give **two** advantages of using polyester for a bottle rather than using glass. (1)

15 Poly(ethene) is a thermoplastic that is used to make ice-cube trays.

a) What happens to thermoplastics when they are heated? (1)

b) From which monomer is poly(ethene) made? (1)

16

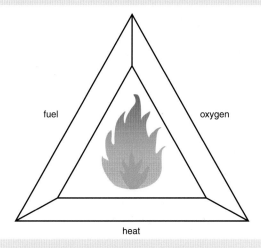

a) Use the fire triangle shown above to explain why a fire blanket can be used successfully to put out the flames of burning clothing on a person. (1)

b) Name the poisonous gas that can be produced when clothing fibres burn in a limited supply of air. (1)

17 The following hydrocarbon substances are found in North Sea oil.

naphtha	23%
kerosene	15%
gas oil	24%
fuel oil	38%

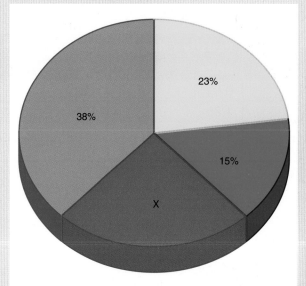

a) Which substance is represented by 'X' in the pie chart above? (1)

b) What name is given to groups of hydrocarbons that have boiling points within a given range? (1)

c) Some naphtha is cracked to give smaller molecules. One reaction that takes place is as follows:

$$C_{10}H_{22} \rightarrow C_7H_{16} + C_3H_6$$

Another reaction is:

$$Y \rightarrow C_6H_{14} + C_2H_4$$

Give the chemical formula for compound Y. (1)

18 Metals used for making aircraft usually have a density of less than 3 density units. They must also have a melting point of more than 600°C.

Some metals and their properties are shown in the table.

metal	density / density units	melting point / °C
A	0.53	181
B	1.85	1551
C	8.90	1410
D	7.87	1535

a) Which metal might be used for making aircraft? (1)

b) Use the information on page 3 of the data booklet to give the name of another metal that has a density of less than 3 density units and a melting point that is above 600°C. (1)

c) A metal has a melting point of –39°C and a boiling point of 357°C. Would it be a solid, a liquid or a gas at room temperature? (1)

19 Natural fibres for clothing can be from plants or animals. Wool and silk come from animals. Cotton and flax come from plants. Present this information as a table with two headings. (1)

20 Hydrogen is being considered as an alternative fuel for cars instead of petrol. Hydrogen produces about three times as much energy as petrol when it burns.

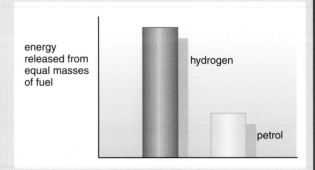

energy released from equal masses of fuel

hydrogen

petrol

a) Suggest a reason why filling a car up with hydrogen could be more difficult than filling it with petrol. (1)

b) When passed through a catalytic converter in a car exhaust system, carbon monoxide and nitrogen oxides react to give two less harmful gases. One of these is nitrogen. Name the other gas that is formed. (1)

c) At 300°C about 90% of the harmful gases from a car engine react to produce less harmful ones when they pass through the catalytic converter. On short journeys, however, most gases pass through the converter unchanged. Suggest a reason for this. (1)

Total 30 marks

Unit 2 Glossary of Terms

alloy   A substance that is a mixture of metals, or of metals and non-metals. Examples include brass, solder and stainless steel.

anodising   A process that increases the thickness of the oxide layer on aluminium (to provide protection against corrosion).

battery   Batteries produce electricity from chemical reactions. They convert chemical energy into electrical energy.

biodegradable   Able to be broken down by bacteria in the soil and rot away.

biogas   A mixture of gases formed by the decomposition of plant or animal material. Consists mainly of methane.

Biopol   A biodegradable plastic.

burning   A chemical reaction during which a substance combines with oxygen, producing heat and light.

combustion   Another word for burning.

corrosion   A chemical reaction that involves the surface of a metal changing from an element to a compound.

cracking   An industrial method of breaking up larger hydrocarbon molecules to produce smaller, more useful molecules.

distillation   A process of separation based on differences in boiling points. The changes of state involved are: liquid → gas → liquid

dry-cleaning   A process using special solvents that are particularly good at dissolving oil and grease stains.

durable   Long lasting, hard wearing.

dyes   Coloured compounds that are used to give bright colours to clothing.

electroplating   A process by means of which a layer of metal is deposited on a substance using electricity. The object is used as the negative electrode in a solution containing ions of the metal being deposited.

fibres   Thin strands used to make, among other things, clothing fabrics.

finite resources   Ones that there are only limited supplies of and that cannot be replaced.

fire triangle   A way of understanding how fires can be put out. This can be done by removing heat or fuel. The fire will also go out if oxygen cannot reach it.

fossil fuels   Fuels formed from plant and animal material over a very long time. Examples include coal, oil, natural gas and peat.

fraction   A group of compounds with boiling points within a given range. Petrol, kerosene and diesel are examples of fractions.

fractional distillation   The process used to separate crude oil into fractions according to the boiling points of the components of the fractions.

fuel   A substance that is burned to produce heat energy.

fuel crisis   Having not enough supplies of fuel to meet the demand for it.

galvanising   A process by which iron is coated with a protective layer of zinc (by dipping into molten zinc).

hydrocarbon   A compound that contains hydrogen and carbon only. An example is methane, CH_4.

incinerate   A process of burning refuse, such as waste plastic, so that it is reduced to ashes.

malleability   The ability to be shaped by hammering, rolling and bending. A physical property of metals.

monomer   A small molecule from which large polymer molecules can be made.

natural fibres   Ones that come from plants and animals. Examples include wool, silk and cotton.

plastics   Synthetic materials made up of polymers.

pollutant   Something that harms the environment.

polymer   A very large molecule that is formed by the joining together of many small molecules called monomers.

polymerisation   The process of joining together many small monomer molecules to make a polymer.

recycle   Convert waste to re-usable material. For example, waste polythene shopping bags are converted into bin bags.

renewable resource   Resources that can be replaced. Examples of renewable sources of energy include methane, ethanol and hydrogen.

rust indicator   A pale yellow solution, which turns blue if rusting is taking place.

rusting   The corrosion of iron. It is caused by the reaction of iron with oxygen and water.

silicones   Compounds used for water-proofing fabrics.

scum   A solid formed by the reaction of some soaps with hard water.

synthetic   Man-made.

synthetic fibres   Ones that are made by the chemical industry. Examples include nylon and polyesters such as Terylene.

thermoplastic   Plastic that softens on heating and can be reshaped.

thermosetting plastic   Plastic that does not soften on heating and cannot be reshaped.

toxic   Poisonous.

Unit 3

Chemistry and Life

The topics covered in this unit are
Photosynthesis and respiration
Effects of chemical growth on plants
Food and diet
Drugs

9 Photosynthesis and respiration

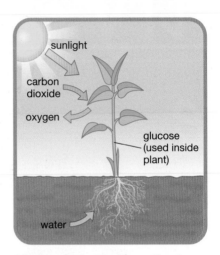

Figure 9.1 *Photosynthesis*

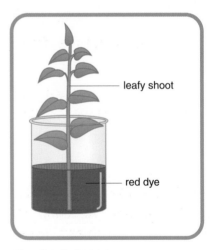

Figure 9.2 *Water passes up through plants*

Photosynthesis

Plants make their own food by taking in substances from the environment. They do this by a process called **photosynthesis** (photo = *light*, synthesis = *making something from simple things*). To do this, plants need water and carbon dioxide from the environment, light and a special green substance called **chlorophyll**. The chlorophyll is found in the leaves of plants. It absorbs the light energy needed for photosynthesis. In the leaves, water and carbon dioxide react to produce glucose and oxygen (see figure 9.1):

carbon dioxide + water → glucose + oxygen

Some glucose is used to provide the plant with energy. The rest of the glucose is used to make starch and cellulose. Starch is a food store for the plant. Cellulose is used to make cell walls, and is the structural material of plants.

We can show that water can move up through a plant stem by putting a leafy shoot into water containing a red dye (see figure 9.2). After a while, the stem can be cut to show that the red dye solution has moved up through it.

In a plant, water is drawn up from the soil by the roots.

Requirements for photosynthesis

Investigating if light is needed

If starch is present in a leaf, it is because photosynthesis has taken place. We can test for the presence of starch using iodine solution. If starch is present in a leaf, the leaf turns blue-black if iodine solution is placed on it. (You will learn more about this test on page 160.)

A plant is left in darkness for two days. Iodine solution is then put onto a leaf from the plant. No blue-black colour is produced, showing that no starch is present in the leaves. Testing a similar plant that has not been kept in a dark place does give a positive test for starch. This experiment shows that light is needed for photosynthesis to take place.

Investigating if carbon dioxide is needed

Two plants that have been kept in darkness for two days are placed under bell jars as shown in figure 9.3. They therefore have no starch in their leaves.

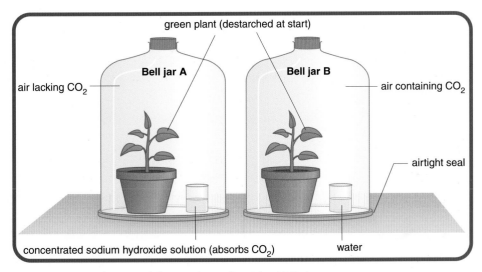

Figure 9.3 *The need for carbon dioxide (CO_2)*

The bell jars are left in bright light for two days and then leaves from each are tested for starch.

Leaves from bell jar **B** give a positive test for starch. Leaves from bell jar **A** have no starch in them. We can therefore conclude that carbon dioxide is needed for photosynthesis.

Investigating if chlorophyll is needed

The leaves of some plants are not all green. Some, like the leaves of a 'wandering sailor' plant, have green and white stripes. When the test for starch is applied to one of these leaves, only the green part turns blue-black (see figure 9.4). This shows that chlorophyll, the green substance in plant leaves, is needed for photosynthesis.

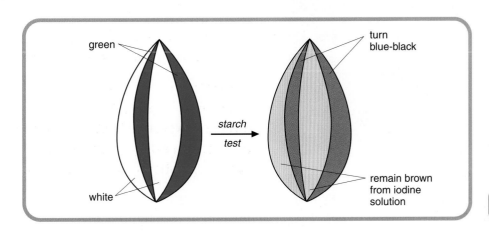

Figure 9.4 *The need for chlorophyll*

Oxygen is a by-product of photosynthesis

The gas given off by the waterweed in figure 9.5 is found to relight a glowing splint. This experiment shows that oxygen gas is formed during photosynthesis.

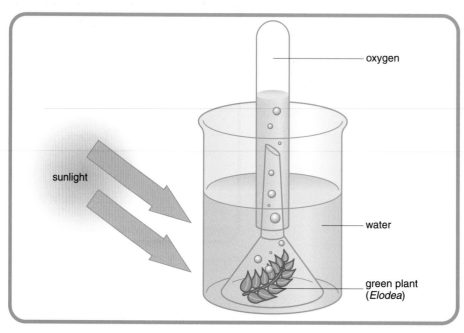

Figure 9.5 *Oxygen is produced during photosynthesis*

Section Questions

1 During photosynthesis,
   a) name the gas taken in by a plant,
   b) name the gas given out,
   c) name the green substance in the leaves that is used to absorb energy.

Respiration

Animals obtain energy from food by a process called **respiration**. Respiration is the reverse of photosynthesis. During respiration, glucose reacts with oxygen to produce carbon dioxide and water:

glucose + oxygen → carbon dioxide + water

Animals obtain glucose by eating food that has come from plants. The reaction above releases energy, which is used by animals in several ways. Two important uses for this energy are for warmth and movement (see figure 9.6).

Figure 9.6 *Glucose provides athletes with the energy they need to run (and to jump)*

Investigating the air we breathe out

The respiration reaction between glucose and oxygen takes place in the cells of our body. We breathe out the water vapour and carbon dioxide that are the products of the reaction.

If we blow through a straw into lime water, it turns milky (see figure 9.7). This shows that there is carbon dioxide in our breath.

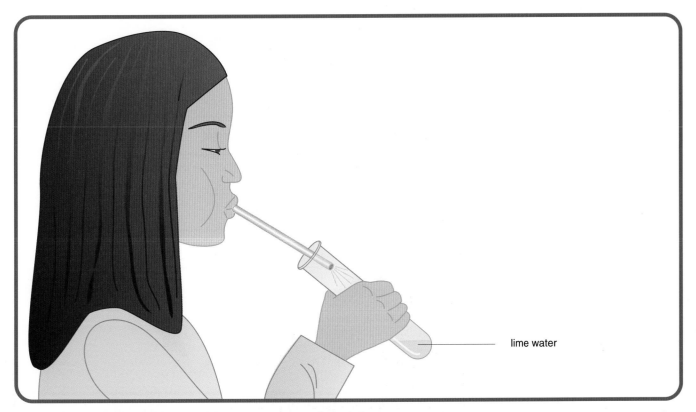

lime water

Figure 9.7 *We breathe out carbon dioxide produced during respiration*

You will have noticed that your moist breath forms a mist as you breathe out on a cold day. You can also show that we breathe out water vapour by breathing onto a cold surface such as a mirror (see figure 9.8).

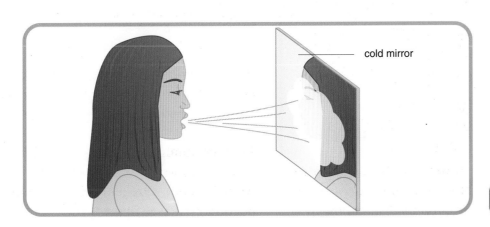

cold mirror

Figure 9.8 *Water condenses on a cold mirror when we breathe on it*

balancing act!

Plants take in carbon dioxide during photosynthesis and give out oxygen. Animals take in oxygen, so that respiration can take place, and give out carbon dioxide. These processes almost balance one another (see figure 9.9). As a result, the amounts of oxygen and carbon dioxide in the air are almost constant.

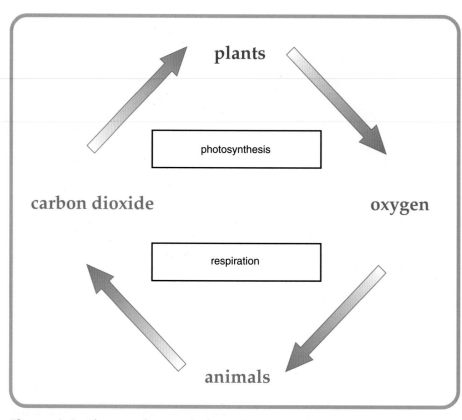

Figure 9.9 *There is almost a balance between photosynthesis and respiration*

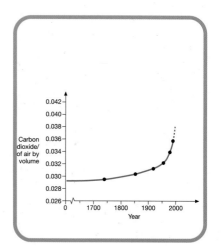

Figure 9.10 *Increasing levels of carbon dioxide in the air*

Section Questions

2 Oxygen and carbon dioxide are involved in the respiration process.
a) Which gas is given out?
b) Which gas is taken in?

3 a) Describe the chemical test for oxygen.
b) Describe the chemical test for carbon dioxide.

The greenhouse effect

Unfortunately, in recent years, carbon dioxide levels in the air have begun to rise (see figure 9.10).

The balance between oxygen and carbon dioxide levels is no longer being maintained. One of the main reasons for this is the cutting down of forests (see figure 9.11). This reduces the amount of carbon dioxide that can be removed from the air during photosynthesis.

Figure 9.11 *Around the world, forests are being cut down*

The burning of fuels increases the amount of carbon dioxide in the air (see figure 9.12). For example, the burning of petrol, diesel, natural gas and coal all produce carbon dioxide.

Figure 9.12 *Burning fuels produce carbon dioxide, which is a 'greenhouse' gas*

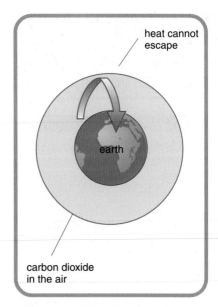

Figure 9.13 *The greenhouse effect*

The carbon dioxide in the air helps to keep the Earth warm. It does this by preventing too much heat being lost from the Earth's surface. Carbon dioxide acts as a sort of blanket around the Earth (see figure 9.13). We call this trapping of heat energy 'the greenhouse effect'. If there were no carbon dioxide in the atmosphere, our world would be a much colder place.

There is evidence that the temperatures around the world are rising steadily. This may well be due to the rise in carbon dioxide levels and is known as **global warming**.

Global warming – an investigation

Does carbon dioxide trap the sun's energy and cause global warming? We can investigate this using the apparatus shown in figure 9.14.

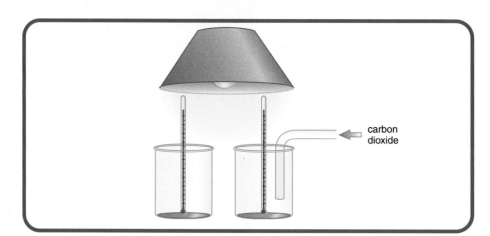

Figure 9.14 *An experiment to investigate global warming*

A bright light is used to act as the sun. Lead foil at the bottom of the beakers takes the place of the surface of the Earth. A slow stream of carbon dioxide is passed into one beaker.

After a short while, the temperature in the beaker containing the carbon dioxide is found to be higher than in the other. This shows that carbon dioxide is trapping heat in the beaker.

Section Questions

4 a) What is happening to the levels of carbon dioxide in the atmosphere?
   b) How is this likely to affect the temperature of the atmosphere?
   c) What name is given to this effect?
   d) What effect does the combustion of fossil fuels have on the level of carbon dioxide in the atmosphere?

Effects of chemicals on growth of plants

Using chemicals to save plants

Many people in the world go hungry through lack of food. It is therefore important that crops produce as good a yield as possible. Chemicals can help farmers and gardeners to achieve this.

The yield of healthy crops can be reduced in the following ways.
- Crops can be eaten by pests such as insects (see figure 10.1) and slugs.
- Bacteria and fungi can cause plants to become diseased.
- Weeds can hinder the growth of plants by, for example, using up essential substances in the soil.

Figure 10.1 *In Africa, locusts are pests. They can totally destroy a crop in hours*

Many chemicals are available at garden centres and from agricultural specialists to help keep plants healthy (see table 10.1 and figure 10.2).

chemical	use
pesticides	control pests
fungicides	prevent diseases
herbicides	kill weeds

Table 10.1 *Three types of chemicals used to keep plants healthy*

Figure 10.2 *All of these chemicals are used to save plants*

We should be particularly careful when using pesticides (see figure 10.3). This is because they are **toxic**. If any comes into contact with skin, for example, it should be washed off immediately.

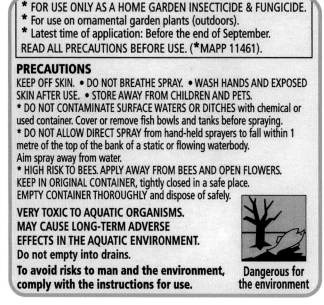

* FOR USE ONLY AS A HOME GARDEN INSECTICIDE & FUNGICIDE.
* For use on ornamental garden plants (outdoors).
* Latest time of application: Before the end of September.
READ ALL PRECAUTIONS BEFORE USE. (*MAPP 11461).

PRECAUTIONS
KEEP OFF SKIN. • DO NOT BREATHE SPRAY. • WASH HANDS AND EXPOSED SKIN AFTER USE. • STORE AWAY FROM CHILDREN AND PETS.
* DO NOT CONTAMINATE SURFACE WATERS OR DITCHES with chemical or used container. Cover or remove fish bowls and tanks before spraying.
* DO NOT ALLOW DIRECT SPRAY from hand-held sprayers to fall within 1 metre of the top of the bank of a static or flowing waterbody.
Aim spray away from water.
* HIGH RISK TO BEES. APPLY AWAY FROM BEES AND OPEN FLOWERS.
KEEP IN ORIGINAL CONTAINER, tightly closed in a safe place.
EMPTY CONTAINER THOROUGHLY and dispose of safely.

VERY TOXIC TO AQUATIC ORGANISMS.
MAY CAUSE LONG-TERM ADVERSE
EFFECTS IN THE AQUATIC ENVIRONMENT.
Do not empty into drains.
To avoid risks to man and the environment,
comply with the instructions for use.

Dangerous for the environment

Figure 10.3 *Care must be taken when using pesticides*

Fortunately, there can often be a more natural way of dealing with pests. Natural predators can be used safely to control them. For example, ladybirds eat greenfly and plant lice (see figure 10.4), but do not harm plants. Gardeners are usually very pleased to see ladybirds.

Figure 10.4 *Ladybirds help to protect crops*

Figure 10.5 *Blue tits feed on caterpillars, thus helping to protect crops*

Small birds can also be excellent natural predators, protecting crops. Blue tits are common garden birds, consuming vast quantities of pests. For example, about 10 000 caterpillars are taken back to feed the young in a single nest (see figure 10.5).

Other natural predators include hedgehogs, which feed on slugs. Spiders can also be useful because they catch insects in their webs.

Section Questions

1 Copy and complete the following:
a) Crop yields can be reduced by
  i)   crops being .... by pests,
  ii) b.............. and f........... causing plants to become diseased,
  iii) weeds inhibiting the .... of plants.
b) Chemicals can save plants.
  i)   .... control pests.
  ii) Fungicides prevent d...............
  iii) Herbicides .... weeds.

2 Give an example of a natural predator and state which pest(s) it controls.

Fertilisers

Figure 10.6 *This plant has received no phosphorus*

Plants take the elements that they need for growth from the soil. The three most important elements for the healthy growth of plants are nitrogen (N), phosphorus (P) and potassium (K). These elements are taken in through the roots of plants as compounds that are in solution. If a plant is deprived of any one of these **essential elements**, it cannot develop properly. Simple experiments can show this (see figures 10.6 to 10.9).

Figure 10.7 *This plant has received no nitrogen*

Figure 10.8 *This plant has received no potassium*

Figure 10.9 *This plant has received phosphorus, nitrogen and potassium*

In areas of natural vegetation, such as woodlands, the decay of vegetable and animal remains returns all essential elements to the soil. However, when crops are harvested, these elements are lost and must be replaced. This is done by the addition of substances called **fertilisers**. Fertilisers contain essential elements for plant growth in the form of suitable compounds. There are two main types of fertiliser.

Natural fertilisers

These are made from animal or plant waste. Animal manure (see figure 10.10) is a natural fertiliser, as is garden compost.

Figure 10.10 *Animal manure is spread on fields as a natural fertiliser*

Artificial fertilisers

These are made by the chemical industry. They are needed because of the increased demand for food. There is not enough natural fertiliser to meet this demand.

To be effective, fertilisers must be soluble in water (see figure 10.11). This is so that they can be taken up by the roots of the plants.

fertilizer

Figure 10.11 *A fertiliser must be soluble in water*

The major artificial fertilisers are soluble ionic compounds. Commonly used ones include ammonium phosphate and potassium nitrate. Table 10.2 shows the commonly used artificial fertilisers.

element needed by plants	compounds present in fertiliser
potassium	potassium compounds
phosphorus	phosphate compounds
nitrogen	ammonium compounds and nitrate compounds

Table 10.2 *Elements needed in artificial fertilisers*

Figure 10.12 *The NPK content is clearly shown on fertiliser bags*

The nitrogen, phosphorus and potassium content of a fertiliser is shown on the bag or packet. This is called the NPK content (see figure 10.12), and is the percentage, by weight, of the elements in the fertiliser.

Section Questions

3 a) The NPK content of a fertiliser is shown on the bag. Give the *names* of these three essential elements.
  b) What is the effect on the growth of a plant if it is deprived of one of these essential elements?

4 Name a *natural* fertiliser.

5 Potassium nitrate and calcium phosphate have both been used as fertilisers.
  a) Refer to page 4 of the data booklet and state the solubility of each.
  b) Which is likely to be the more effective fertiliser and why?

Prescribed Practical Activity

Solubility

Information

The aim of the experiment is to test the solubility in water of some ammonium, nitrate and phosphate compounds. This is to decide if they could be used as fertilisers. Only the ones that dissolve in water will be able to fertilise plants.

What to do

1 Fill a beaker half full of water.

2 Pour water into a test tube to a depth of 3–4 cm.

3 Using a spatula, add a tiny amount of ammonium sulphate to the water.

4 Hold the test tube at the mouth and shake it gently for several minutes as in figure 10.13.

5 Look at the mixture to see if it has dissolved.

6 Record your results, giving the name of the compound and whether it was soluble or not.

7 Repeat the experiment with each of the remaining compounds.

8 Record your results.

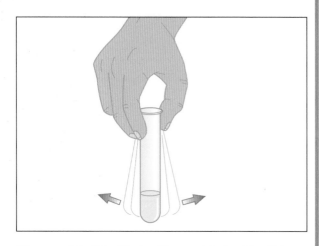

Figure 10.13 *Shake the test tube like this*

Results table

name of compound	soluble / insoluble
ammonium sulphate	soluble
potassium nitrate	soluble
sodium nitrate	soluble
calcium phosphate	insoluble
ammonium phosphate	soluble

Conclusion

All of the compounds in the table could act as fertilisers, except calcium phosphate. This is because calcium phosphate is insoluble in water.

Figure 10.14 *Bottled water is very low in nitrates*

Nitrate pollution

In recent years there has been a large increase in the use of artificial nitrate fertilisers. These are very soluble and are easily washed out of the soil. Increased levels of nitrates have been noted in rivers, in lochs and in public water supplies. This may be due to the use of artificial fertilisers. However, we cannot be certain, because natural fertilisers, like manure, contain nitrates as well.

It is believed that high levels of nitrates in drinking water are harmful, particularly to young children. In some areas, infants are given bottled water as a precaution. Most bottled water has very low nitrate levels (see figure 10.14 and table 10.3).

For example, Buxton Water shows very low nitrate levels.

typical analysis	mg/litre
calcium	55
magnesium	19
potassium	1
sodium	24
bicarbonate	248
chloride	37
sulphate	13
nitrate	less than 0.1

Table 10.3 *Official analysis of Buxton Water*

The presence of large quantities of nitrates can leave lakes, rivers, seas and oceans lifeless. This is because the growth of simple plants called **algae** is boosted. The algae grow quickly and form a layer of 'algal bloom' on the surface of the water (see figure 10.15).

Figure 10.15 *An algal bloom*

When the algae die and decompose, they remove dissolved oxygen from the water. Without oxygen, animal and plant life cannot survive, and the water becomes lifeless.

Section Questions

6 a) What is a likely cause of nitrate pollution in water supplies?
   b) Explain how algal blooms, produced by nitrate pollution, can cause water to become lifeless.

Root nodules

Some plants have small lumps on their roots called **root nodules**. They can be seen on the roots of the clover plant in figure 10.16.

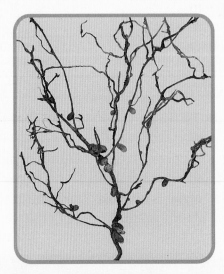

Figure 10.16 *Can you see the root nodules?*

Inside the root nodules, nitrogen from the air is converted into nitrates. Plants like clover thus make their own fertiliser! Other plants that have root nodules include peas and beans. These can also convert nitrogen from the air into nitrates.

Farmers can use plants that have root nodules to increase the fertility of soil. For example, a field can be planted with clover. Once the clover plants have grown, they are then ploughed back into the soil. As they decay, the clover plants release nitrates into the soil to act as fertilisers for other crops.

Section Questions

7 a) Name three plants that possess root nodules.
  b) Which gas from the air is converted into nitrates inside root nodules?

ACCESS 3 Subsection Test: Photosynthesis and respiration, and Effects of chemicals on growth of plants

Part A

This part of the paper consists of four questions and is worth 4 marks.

1 Which gas reacts with glucose during respiration? (1)

    oxygen or **carbon dioxide**

2 What do plants make during photosynthesis? (1)

    carbon dioxide or **glucose**

3 Which of the following is a 'greenhouse gas' and is believed to be responsible for global warming? (1)

    nitrogen or **carbon dioxide**

4 Which of the following prevents diseases in plants? (1)

    fungicides or **herbicides**

Part B

This part of the paper is worth 6 marks.

5 The information in the diagram refers to the mass of fertilisers containing nitrogen that have been used in the past.

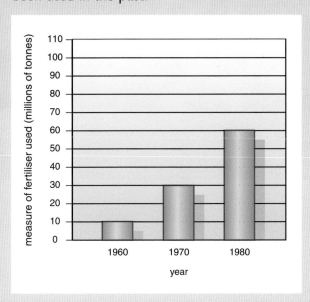

a) If the same trend continued, predict the mass of fertiliser (in millions of tonnes) that was used in 1990. (1)

b) Give the name of another element, in addition to nitrogen, that is essential for healthy plant growth. (1)

6 The graph shows how the amount of carbon dioxide in the air has changed.

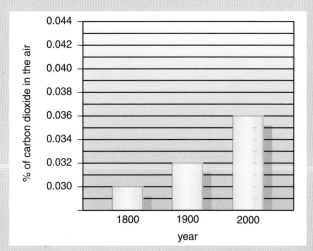

a) What was the percentage of carbon dioxide in the air in the year 2000? (1)

b) What has happened to the level of carbon dioxide in the air over the last two hundred years? (1)

7 Pesticides can be used to control pests that eat crops. Why must care be taken when using pesticides? (1)

8 Peter tested the solubility of three compounds that have been tried as fertilisers. He found that ammonium nitrate was very soluble, that urea was soluble, but that calcium phosphate was insoluble.

Give the results of Peter's experiments as a table using the two headings 'compound' and 'solubility'. (1)

Total 10 marks

11 Food and diet

Elements in the human body

A **balanced diet** is one that provides the body with all the essential elements and compounds that our bodies need. Figure 11.1 shows some of the foods that can make up a healthy diet. How many of the foods shown do you eat regularly?

Figure 11.1 *Foods for a healthy diet*

Elements are present in our diet and in our bodies as chemical compounds, not as free uncombined elements.

The main compounds that are essential in our diet are **carbohydrates**, **fats** and **proteins**.

The main elements in our bodies are carbon, hydrogen, oxygen and nitrogen. Together, carbohydrates, fats and proteins provide these four basic elements that we need (see table 11.1).

compounds	elements present
carbohydrates	C, H and O
fats	C, H and O
proteins	C, H, O, N and other elements

Table 11.1 *The main food compounds in our diet*

Our bodies also need small quantities of **minerals**, which are vital for a healthy body. Iron, for example, is needed to make the red compound in blood. Calcium is necessary for healthy bones and teeth. The body also needs very small amounts of **trace elements** such as zinc and copper. Although important, some trace elements are toxic if taken in too large quantities and can therefore poison us.

Water is very important in our diet. Water makes up more than 60% of our body weight. Although it has no food value, we cannot live for many days without water. If you take a holiday in a hot country, you will find that you need to drink a lot of water.

We are sometimes uncertain about what we should eat. Which foods contain the elements and compounds that we need? Table 11.2 will help you to decide.

element or compound	sources
carbohydrate	bread, pasta, potatoes, cakes, biscuits, sugar
fat	butter, margarine, cream, cheese, cooking fat and oil
protein	meat, fish, eggs, peas, beans, lentils
calcium	milk, cheese, yoghurt
iron	red meat, liver, beans, lentils, green vegetables

Table 11.2 *Some sources of the elements and compounds that our bodies need*

Section Questions

1 Name the four main elements in our bodies.

2 Name the most common compound in our bodies.

3 Proteins are essential compounds in our diet. Name two others.

4 a) Why do we need calcium in our diet?
   b) Why do we need iron in our diet?

Carbohydrates

Carbohydrates are an important class of food compounds, which are made by plants. They are made by the process called **photosynthesis**, which was described on page 142. Carbohydrates can be divided into **sugars**, which are sweet, and **starch**, which is not. Examples of sugars include glucose, sucrose (table sugar), fructose and maltose.

carbohydrates

sugars   starch

Testing for starch

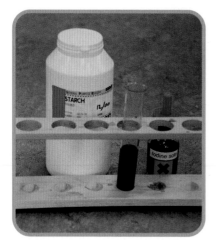

Figure 11.2 *Adding iodine solution to starch produces a dark blue (blue-black) colour*

Starch is the only carbohydrate that gives a dark blue colour (sometimes almost black) with **iodine solution**. Figure 11.2 shows the result of testing a solution of starch with iodine solution.

Bread, potatoes and pasta all contain starch, the presence of which can be shown using iodine solution.

Elements in carbohydrates

carbohydrate	formula
glucose	$C_6H_{12}O_6$
fructose	$C_6H_{12}O_6$
sucrose	$C_{12}H_{22}O_{11}$
maltose	$C_{12}H_{22}O_{11}$

Table 11.3 *Chemical formulae for some carbohydrates*

All carbohydrates contain the elements carbon, hydrogen and oxygen. The chemical formulae for some are given in table 11.3.

An interesting way of showing that the element carbon is present in carbohydrates is to add concentrated sulphuric acid. This removes all of the hydrogen and oxygen in the carbohydrates to leave a black mass of carbon (see figure 11.3).

Figure 11.3 *Adding concentrated sulphuric acid to a carbohydrate*

Dissolving carbohydrates in water

All of the sugars dissolve easily in cold water, but starch does not (see table 11.4). Very hot water needs to be added to starch in order to dissolve it.

carbohydrate	solubility in cold water
glucose	soluble
sucrose	soluble
fructose	soluble
maltose	soluble
starch	insoluble

Table 11.4 *Sugars dissolve readily in water, but starch does not*

Reactions of carbohydrates

Making starch from glucose

Small sugar molecules, like glucose, are made by plants during photosynthesis. Inside the plant, small glucose molecules join together to form much larger starch molecules. Plants convert the glucose into starch for storing energy. The starch can be converted back to glucose, for use in the process of respiration, when energy is needed.

Glucose molecules are monomers. They can be converted into the polymer, starch, by the reaction known as polymerisation:

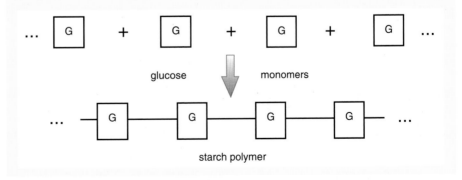

Energy from carbohydrates

Within our bodies, carbohydrates provide us with energy as a result of the respiration reaction:

glucose + oxygen → carbon dioxide + water

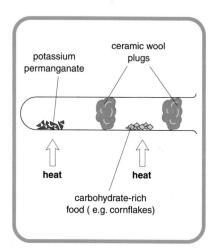

Figure 11.4 *Carbohydrates burn very well when heated in oxygen*

Various experiments can be carried out to show that reactions between carbohydrates and oxygen release energy. In one experiment, shown in figure 11.4, a carbohydrate food, such as cornflakes, can be made to burn brilliantly in the oxygen released from potassium permanganate crystals.

In another experiment, a powdered carbohydrate can be made to burn explosively when it is injected into a tin, as shown in figure 11.5. The lid is usually blown out of the tin when the powdered carbohydrate catches fire.

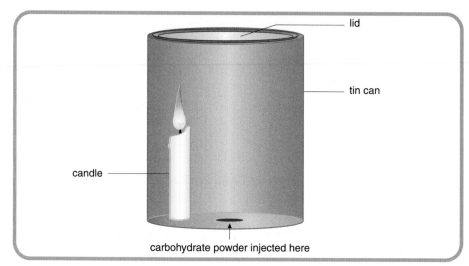

Figure 11.5 *Carbohydrate powder can burn explosively in this experiment*

Prescribed Practical Activity

Burning carbohydrates

Information

The aim of the experiment is to show that heat energy is produced when sugar and starch are burned, and to compare how much heat energy each produces.

Flour is burned as the 'starch' carbohydrate and icing sugar as the 'sugar' carbohydrate (see figure 11.6). The temperature rise of 10 cm^3 of water is used to give an idea of how much heat energy is produced by the burning of each carbohydrate.

Figure 11.6 *These foods provide us with the carbohydrates starch and sugar*

What to do

1 Half-fill a beaker with water.
2 Add water to a measuring cylinder up to the 10 cm³ mark.
3 Pour this 10 cm³ of water into a boiling tube.
4 Clamp the boiling tube in a vertical position.
5 Measure the temperature of the water in the boiling tube and record the result.
6 Light a bunsen burner and add flour to a spatula to give a level spatulaful.
7 Heat the flour in the bunsen flame until it just catches fire.
8 Quickly place the burning flour underneath the boiling tube so that the flames are touching the bottom of the boiling tube, as in figure 11.7.
9 When the flour has stopped burning, stir the water gently with the thermometer. Measure and record the final water temperature.
10 Repeat the experiment using icing sugar. To make the experiment as fair as possible, use a level spatulaful of icing sugar, just as you did for the flour.

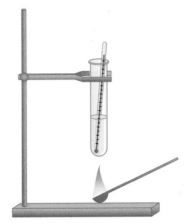

Figure 11.7 *Heating water with a burning carbohydrate*

Results table

carbohydrate	starting temperature of water / °C	final temperature of water / °C	rise in temperature / °C
flour (starch)	20	33	13
icing sugar (sugar)	20	36	16

Conclusion

Sugar releases slightly more energy than starch when equal masses are burned.

Digesting starch

Before our bodies can use the energy stored in starch, it must first be broken down into glucose. This happens as we digest our food. The starch, in such foods as bread and potatoes, is broken down into glucose. Unlike starch, glucose is soluble and small enough to pass through the gut wall and into the bloodstream. Once in the blood, it can be taken to body cells, where respiration takes place, releasing energy.

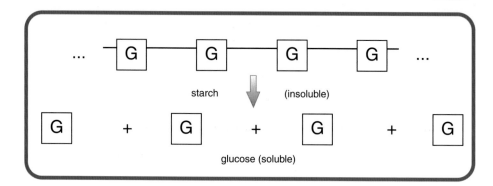

In the laboratory, starch solution can be broken down using either the body enzyme called amylase or dilute acid (see figure 11.8). If amylase is used, the mixture is kept at about 37°C (body temperature). If acid is used, the mixture is heated to about 100°C.

Testing with iodine solution and Benedict's solution shows that the starch is broken down to a sugar, possibly glucose (see table 11.5). This happens with both mixtures.

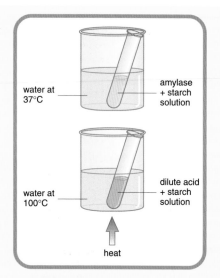

Figure 11.8 *Two ways of breaking down starch*

test solution	at start	after 15 minutes
iodine	dark blue	no effect
Benedict's	no effect	orange-red

Table 11.5 *Test results*

Amylase is one of many enzymes present in our bodies. As figure 11.9 shows, it works best at 37°C, which is normal body temperature. Unfortunately, body enzymes are destroyed at higher temperatures, as can be seen from the graph.

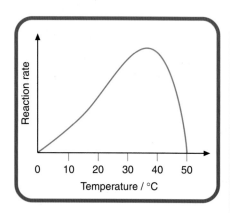

Figure 11.9 *How body enzyme activity changes with temperature*

Section Questions

5 What are carbohydrates used for in the body?

6 Give two examples of sugars.

7 What colour is produced when iodine solution is added to starch?

8 Which sugar is not detected by the Benedict's test?

9 a) Which sugar is starch broken down to during digestion?
   b) Starch can be broken down by the enzyme amylase. What else can break down starch?
   c) Describe how you would show that the body enzyme amylase is destroyed by boiling water.

Testing for starch and sugars in food

Information

The aim of the experiment is to test for starch and sugars in some food samples.

Iodine solution (see figure 11.10) is used to test for starch, with which it produces a blue-black colour.

Benedict's solution (see figure 11.10) is used to test for sugars, with which it produces an orange-red colour when heated. Glucose, fructose, maltose and some other sugars react in this way. Sucrose, however, does not react with Benedict's solution.

What to do

Testing for starch

1 Add small samples of milk, egg white, bread and potato to separate dimples in a dimple tray.
2 To each sample, add a few drops of iodine solution.
3 Observe what happens.
4 Record your results.

Figure 11.10 *These solutions are used to test for starch and sugars*

What to do

Testing for sugars

1 Boil some water in a kettle and half fill a beaker with the very hot water.
2 Add some milk to one test tube, to a depth of about 1 cm.
3 Add some egg white to another test tube, to a depth of about 1 cm.
4 Add bread crumbs to a third test tube.
5 Add small pieces of potato to a fourth test tube.
6 To all four test tubes, add Benedict's solution to a depth of about 3 cm (see figure 11.11).
7 Place the test tubes in the hot water and observe what happens.
8 Record your results.

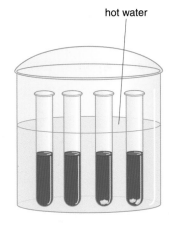

hot water

Figure 11.11 *Testing for sugars with Benedict's solution*

Results table

food	observations on adding iodine solution	observations on adding Benedict's solution
milk	no change	orange-red colour
egg white	no change	no change
bread	blue-black colour	no change
potato	blue-black colour	no change

Conclusions

food	Is starch present?	Is sugar present?
milk	no	yes
egg white	no	no
bread	yes	no
potato	yes	no

Fats and oils

Fats and oils are an important class of food compounds. **Fats** are mainly obtained from animals. **Oils** are mainly obtained from plants.

Fats and oils are very similar in molecular structure. However, fats are solids at room temperature, whereas oils are liquids. Supermarket shelves contain a lot of fats and oils (see figures 11.12 and 11.13).

Figure 11.12 *Some common fats*

Figure 11.13 *Some common oils*

Table 11.6 shows the names of some common animal fats and vegetable oils.

fats	oils
lard (pigs)	olive
dripping (cattle)	corn
mutton fat (sheep)	rapeseed
	sunflower

Table 11.6 *Some common fats and oils*

Some of our foods contain a lot of fats and oils. Some typical values are given in table 11.7.

food	fat or oil / g per 100 g
butter	83
peanuts	49
cheese	35
sausages	31
cake	16
eggs	12

Table 11.7 *Fat or oil contents of some foods*

Energy from fats and oils

One gram of fat or oil can produce more than twice as much energy as one gram of carbohydrate (see figure 11.14).

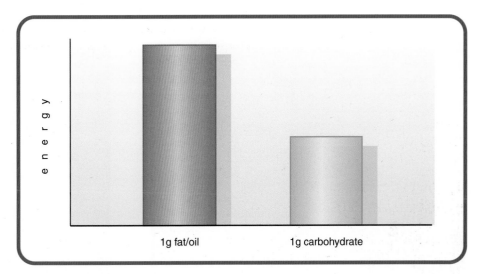

Figure 11.14 *Fats and oils are a much more concentrated energy source than carbohydrates*

The amounts of carbohydrate and fat present in food are shown on the food label. Have you noticed that there is also information about how much energy the food gives us. Table 11.8 shows the information given on some common foods.

food	carbohydrate / g per 100 g	fat / g per 100 g	energy / kJ per 100 g
butter	0.1	81.5	3050
peanuts	7.8	53.0	2580
crisps	30.5	12.0	2035
bread	45.9	0.8	970

Table 11.8 *Foods high in fat provide a lot of energy*

In our bodies, fats and oils provide us with energy by reaction with oxygen. In this way, fats and oils are like carbohydrates. Just like carbohydrates, it is possible to burn fats and oils, releasing energy much quicker than in the body. The presence of a wick, such as a piece of pipe cleaner, helps fats and oils to burn (see figure 11.15). It is interesting to discover that peanuts, which are high in fat/oil content, can provide enough energy to boil water in a test tube.

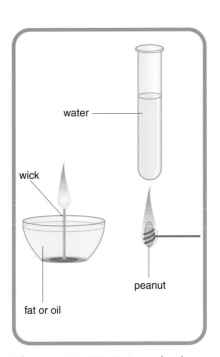

water

wick

peanut

fat or oil

Figure 11.15 *Fats and oils can burn*

Testing for fats and oils

Fats and oils leave a **greasy mark** on filter paper. This is also true of foods that contain fats or oils. Sausage rolls, for example, leave a greasy mark on filter paper (see figure 11.16).

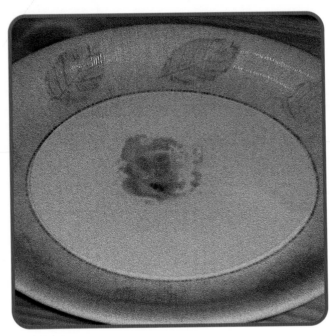

Figure 11.16 *Proving that sausage rolls contain fat (by the greasy mark left behind)*

Saturated fat and heart disease

Figure 11.17 *Sunflower oil is looked upon as healthy food*

There are several kinds of fats and oils, but scientists have identified two types that are particularly important to us. One type is **saturates** and the other is **polyunsaturates**.

- Saturates are believed to increase the cholesterol level in the blood, and this in turn may cause heart disease.
- Polyunsaturates are considered to be potentially less harmful to the heart.

It is also recognised that many people have too much fat in their diet.

As a result, scientists recommend that we should do two things:
- Lower the total fat intake in our diet.
- Replace some of the saturates in our diet with polyunsaturates.

Sunflower oil and margarine made from sunflower oil are particularly high in polyunsaturates (see figure 11.17).

Coronary heart disease

The United Kingdom has one of the worst records for deaths by coronary heart disease anywhere in the world. Causes of death in the UK for 1998 are shown in Figure 11.18.

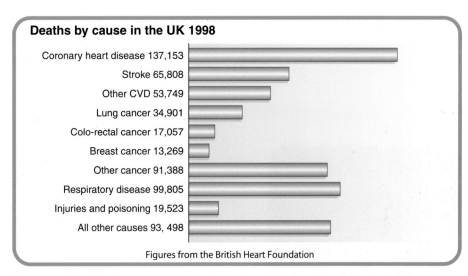

Deaths by cause in the UK 1998

Coronary heart disease 137,153
Stroke 65,808
Other CVD 53,749
Lung cancer 34,901
Colo-rectal cancer 17,057
Breast cancer 13,269
Other cancer 91,388
Respiratory disease 99,805
Injuries and poisoning 19,523
All other causes 93, 498

Figures from the British Heart Foundation

Figure 11.18 *Too many people in the UK die prematurely from heart disease*

Section Questions

10 How does the energy provided by one gram of fat compare with that provided by one gram of carbohydrate?

11 Refer to page 7 of the data booklet and give the mass of fat in grams present in 100 grams of each of the following foods:
a) eggs,
b) jam,
c) steak.

12 Why should you replace some of the saturated fat (saturates) in your diet with polyunsaturated oils (polyunsaturates)?

Proteins

Proteins are another important class of food compounds. They can be obtained from both plants and animals. Proteins are very important to the body, since they provide material for body growth and repair. Some foods that act as important sources of proteins are shown in figure 11.19.

Figure 11.19 *These foods are important sources of protein*

The protein content of some foods is shown in Table 11.9.

food	protein / g per 100 g
peanuts	28
cheese	25
steak	17
fish	16
eggs	12
peas	6
milk	3

Table 11.9 *Protein content of some foods*

Figure 11.20 *Hair contains the protein keratin*

It is said that 'we are what we eat', and this is particularly true of proteins. Important parts of our bodies are made of protein. Muscles, hair, nails, bones, tendons and skin all contain proteins. In our bodies, the proteins we eat are made into the particular proteins that we need. This is true of all animals. For example, hair, nails and skin all contain the protein keratin (see figure 11.20). The most common body protein, collagen, is found in tendons and bones.

Proteins and growth

As we grow, there is a need for more proteins (see figure 11.21). This is because our bones, muscles, tendons, etc., become larger and all contain proteins. Also, as hair and nails grow, even more protein is needed.

Figure 11.21 *Proteins are needed for growth*

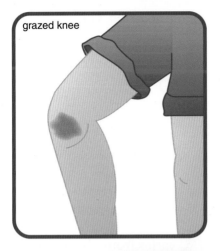

grazed knee

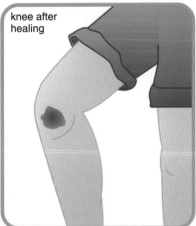

knee after healing

Figure 11.22 *Proteins help in wound healing*

Proteins and tissue repair

Have you ever wondered what really happens when a grazed knee heals (see figure 11.22)? Proteins play a big part in stopping any bleeding, then in scab formation, and finally in the formation of a new layer of skin (which might be scarred). A very important role played by proteins is in fact in wound healing.

Section Questions

13 Proteins are used for body growth. Give two examples of body growth that contain proteins in their structure.

14 a) Apart from body growth, what other important task do proteins provide material for in the body?
    b) Give an example of this use of protein material.

Elements in proteins

Like carbohydrates and fats, proteins are chemical compounds that contain the elements carbon, hydrogen and oxygen. Proteins, however, also contain the element nitrogen (see table 11.10).

compound	elements present
carbohydrate	C, H and O
fat	C, H and O
protein	C, H, O and N

Table 11.10 *Elements present in carbohydrates, fats and proteins*

A chemical test for proteins

The nitrogen that is present in proteins can be released in an alkaline gas when a protein is heated with soda lime. Your teacher may let you heat some hair or nail with soda lime. You should see moist pH paper turn blue-violet at the mouth of the test tube (see figure 11.23).

Proteins are polymers

Proteins are very large molecules and are classified as **polymers**. The monomers that make them up are called **amino acids**. The polymers that we call plastics often contain only one type of monomer.

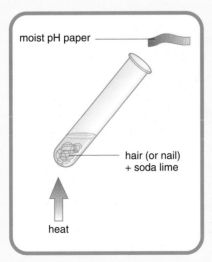

moist pH paper

hair (or nail) + soda lime

heat

Figure 11.23 *Testing for protein*

Proteins, however, contain many different amino acids. The amino acids all look like this:

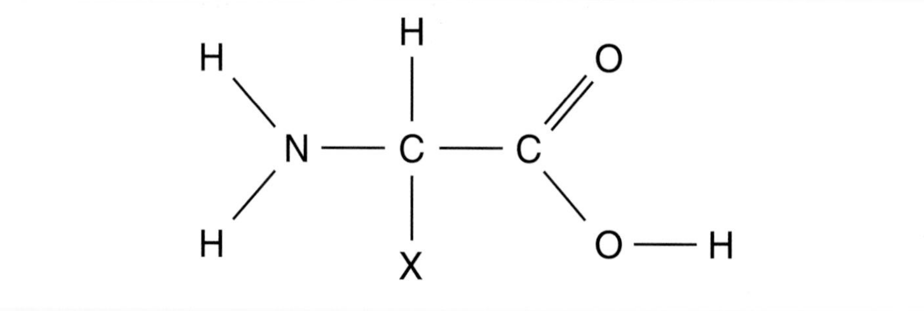

The part of the molecule marked with an 'X' varies from one amino acid to another. In the simplest amino acid, called glycine, 'X' is a hydrogen atom (see figures 11.24 and 11.25).

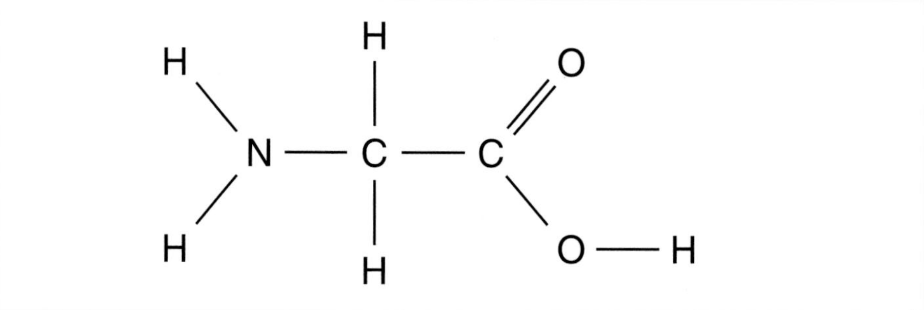

Figure 11.24 *Glycine is the simplest of the amino acids*

Figure 11.25 *A model of a glycine molecule*

Breaking down and building up proteins

When we eat proteins, our bodies break them down into amino acids. This happens during the process called **digestion:**

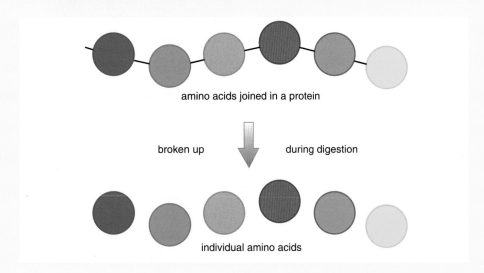

amino acids joined in a protein

broken up   during digestion

individual amino acids

Here the patterned circles represent different amino acids.

The individual amino acids that are produced by the digestion of proteins are soluble. They can be carried by the blood to wherever they are needed in the body. There, they are rebuilt into the various proteins that are needed:

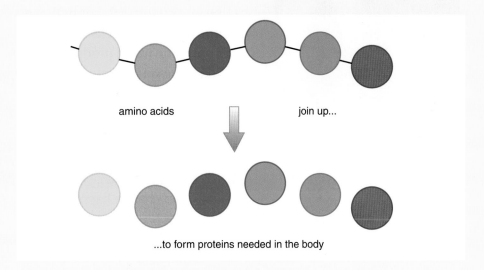

amino acids   join up...

...to form proteins needed in the body

Body proteins may contain different amino acids from the ones in a particular food that we eat. It is therefore important that we eat a variety of foods that contain proteins. For most people, the necessary amino acids are provided from both animal and vegetable sources.

Vegetarians do not eat the flesh of animals. They obtain their proteins only from vegetable sources. They must eat a wide variety of these in order to obtain the necessary amino acids.

Section Questions

15 Which element is present in all protein, but not in carbohydrates or fats?

16 When proteins are heated with soda lime, a gas is given off. Is this gas acidic, neutral or alkaline?

17 Refer to Table 11.9 and name two vegetable foods that are a source of proteins.

18 What is the name of the monomer molecules that are produced when proteins are broken down during digestion?

Fibre

Figure 11.26 *Foods rich in fibre*

After food has been digested and absorbed, waste products must be removed from the body. This process is called excretion, and it is helped by the presence of **fibre** in the diet. Fruit, vegetables and wholegrain bread and cereals are good sources of fibre (see figure 11.26).

Fibre keeps the gut working well. With a high-fibre diet, waste products can be moved through our bodies easily. This is because the fibre absorbs water and swells. The swelling provides bulk for gut muscles to work on as the waste products are squeezed along. In the case of a low-fibre diet, the waste material tends to be less bulky and harder. This makes it more difficult to move along and causes constipation.

Section Questions

19 a) Name two sources of fibre in the diet of humans.
   b) What change happens to fibre in the gut when it absorbs water?
   c) With a low-fibre diet, waste material in the gut is harder, less bulky and more difficult to move along. What name is given to this problem?

Vitamins

Vitamins are a group of complex chemical compounds that are vital to the body. We need them to stay healthy. Lack of important vitamins can cause poor health.

The body requires only small amounts of each vitamin. However, as most are not made in the body, they must be supplied in food.

Vitamins are referred to by letters.
- Vitamins A, D, E and K are fat-soluble.
- Vitamins B and C are water-soluble.

Vitamin B was originally thought to be a single compound. It is now known to contain at least thirteen different ones.

Table 11.11 gives information about the foods in which vitamins A, B, C and D are found, and the effects caused by a lack of them.

vitamin	food found in	effects of lack
A	green vegetables, milk, butter, liver, cod liver oil, margarine	reduced resistance to disease, poor night vision
B	cereal, yeast, meat, eggs	loss of appetite and weight
C	oranges, lemons, tomatoes, fresh vegetables	weakness, bleeding gums, anaemia, illness (scurvy)
D	cod liver oil, egg yolk, cream, margarine	soft bones (leg bones may bend), illness (rickets)

Table 11.11 *Vitamins*

Vitamin E is added to margarine by law. Apart from helping to prevent margarine going rancid, it is believed to reduce the risk of certain cancers. A diet that is high in vitamin C also seems to lower the risk of cancers (see figure 11.27). There is also evidence that the higher the intake of vitamins A and C, the lower the risk of heart disease.

Figure 11.27 *These vitamin C-rich fruit and vegetables help to keep us healthy*

Estimating vitamin C

When iodine solution is added to a solution of vitamin C, the reddish-brown iodine solution is decolourised (see figure 11.28). The volume of iodine solution decolourised is a measure of the vitamin C content of the solution.

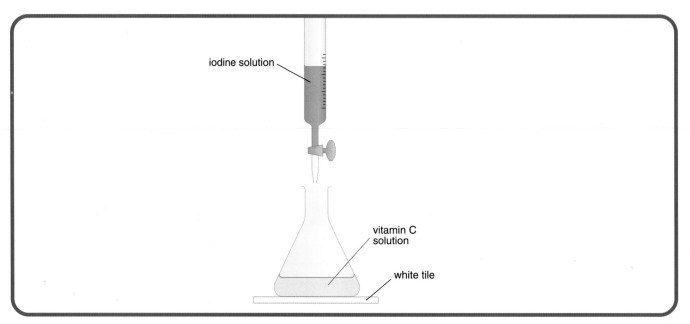

iodine solution

vitamin C solution

white tile

Figure 11.28 *Vitamin C can be estimated using iodine solution*

Section Questions

20 a) Why are vitamins important to the body?
   b) Name three sources of vitamin C.
   c) Name an illness caused by the lack of vitamin D.

21 The vitamin C content of some foods is shown in the table.

food	vitamin C content / mg per 100 g food
apple	20
pear	30
orange	90
cauliflower	100
red pepper	190

Present this information as a bar graph.

Food additives

Figure 11.29 *Cereals contain added vitamins and minerals*

Food additives can be used to supply or enhance the nutritional value of a food. For example, iron is added to breakfast cereals along with various vitamins (see figure 11.29 and table 11.12). Vitamins and minerals, like iron, are added to many foods.

protein	7 g
carbohydrate	84 g
fat	0.8 g
fibre	2.5 g
vitamin B$_1$	1.2 mg
vitamin B$_2$	1.3 mg
vitamin B$_6$	1.7 mg
niacin	15 mg
folic acid	167 µg
vitamin B$_{12}$	0.75 µg
iron	7.9 mg

Table 11.12 *Typical nutrition information (in 100 g of a cereal)*

Some of the quantities added are very small. The symbol 'mg', for example, stands for milligrams, that is thousandths of a gram. The symbol 'µg' means something even smaller – micrograms, or millionths of a gram.

Sometimes there are a lot of additives put into food, as shown in figure 11.30. Fortunately, they can only be used if they have been tested and approved.

Figure 11.30 *This food contains a lot of additives – all tested and approved*

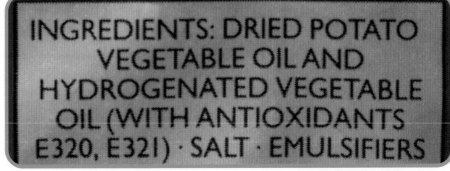

INGREDIENTS: DRIED POTATO VEGETABLE OIL AND HYDROGENATED VEGETABLE OIL (WITH ANTIOXIDANTS E320, E321) · SALT · EMULSIFIERS

Section Questions

22 a) Why are additives put into our foods?
   b) Name two types of additive that are put into foods we eat.

'E' numbers

Food additives are sometimes listed on packaging simply as an 'E' number. This means that the additive has been shown to be safe in the foods in which it is allowed.

Food colours

'E' numbers that begin with a '1' are food colours. Some are obtained from natural sources. For example, E140 is chlorophyll, which is extracted from green cabbage. E162 is betanin, which is the coloured substance found in beetroot. Consumer pressure has resulted in the synthetic colour tartrazine (E102) being removed from many foods. This is because it was believed to affect behaviour.

If more than one coloured substance is present in a food, it is sometimes possible to show this by chromatography. A green colour may, for example, be a mixture of yellow and blue additives (see figure 11.31)

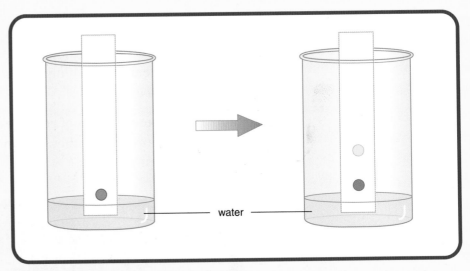

Figure 11.31 *Separating food colours using paper chromatography*

Food flavours

Food flavour additives can be natural or synthetic. The crisps in figure 11.32 contain dried onions and dried cheese to provide flavour.

More than 3500 flavourings are used in the food industry, but at the moment they do not have to be listed on the label. Often all that is said on the label is that the food contains 'flavouring'.

Figure 11.32 *These crisps contain natural flavours*

If a 'flavour enhancer' is used, then it must be listed on the food label. The tin of soup shown in figure 11.33 contains the flavour enhancer monosodium glutamate (E621).

Food preservatives

Food preservatives are added to help prevent the food from going bad. Vinegar, common salt, sugar and wood smoke have been used as food preservatives for centuries. Many other chemicals are now being used for this purpose. Their 'E' numbers all begin with a '2'.

For example, the corned beef in figure 11.34 contains E250, sodium nitrite, as preservative.

Figure 11.33 *This soup contains a flavour enhancer*

Figure 11.34 *Sodium nitrite (E250) helps to preserve this meat*

Figure 11.35 *Sulphur dioxide prevents this fruit from going bad*

Sulphur dioxide (E220) is widely used to preserve fruit juices and dried fruit. The dried apricots in figure 11.35 contain some sulphur dioxide.

Section Questions

23 Give an example of a natural colour that is used as a food additive.

24 What is the name of the flavour enhancer E621?

25 What kind of additives are sodium nitrite and sulphur dioxide?

ACCESS 3 Subsection Test: Food and diet

Part A

This part of the paper consists of four questions and is worth 4 marks.

1 Which carbohydrate is iodine solution used to test for? (1)

   sucrose or **starch**

2 During digestion starch is broken down into: (1)

   glucose or **sucrose**

3 Which fats are thought to be less harmful to the heart? (1)

   saturates or **polyunsaturates**

4 Which is the more important use of proteins in the body? (1)

   to provide energy or **to repair tissue**

Part B

This part of the paper is worth 6 marks.

5 The pie chart shows the main elements in the human body.

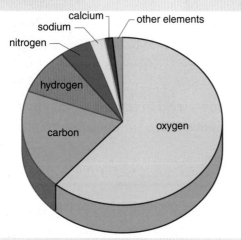

a) Which is the third most plentiful element in the human body? (1)

b) Are these elements present in the body as free elements or are they combined in compounds? (1)

6 Fibre is needed in our diet to keep the gut working well. What problem is caused when waste food material can only be moved slowly through the gut due to lack of fibre? (1)

7 Some peanuts were found to have the following important food compounds present.

food compound	%
fat	50
protein	25
carbohydrate	10

a) Present this information as a bar graph. (1)

b) Explain why peanuts can provide us with a lot of energy. (1)

8 The table shows death rates from heart attack in Ireland from 1983 to 2000 in people up to age 64 years.

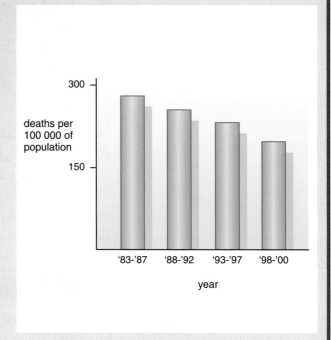

What conclusion about the trend in the rate of deaths from heart attacks in Ireland can you make based on this information? (1)

Total 10 marks

12 Drugs

A **drug** is a substance that alters the way the body works. Many substances can affect the way our bodies work, and are therefore classified as drugs.

Alcohol

Alcohol is classified as a drug because of its effect on the body. Even a small quantity can affect our reaction time and concentration. This is why there is a legal limit for blood alcohol as far as driving is concerned. A high proportion of road accidents involve people driving while over the legal limit for alcohol in their blood (see figure 12.1).

Figure 12.1 *Drinking and driving can have fatal consequences*

When taken in excess, alcohol can have very harmful effects on the body. This is particularly true of the liver and the brain. A badly damaged liver may need to be removed. A transplant operation is then the only hope of survival. Heart muscles can also be weakened by alcohol. This can lead to heart failure.

Units of alcohol

The amount of alcohol in a drink depends upon its volume and the concentration of alcohol in it. The alcohol content of drinks is commonly measured in **units**. Figure 12.2 shows the approximate number of units of alcohol in some common drinks.

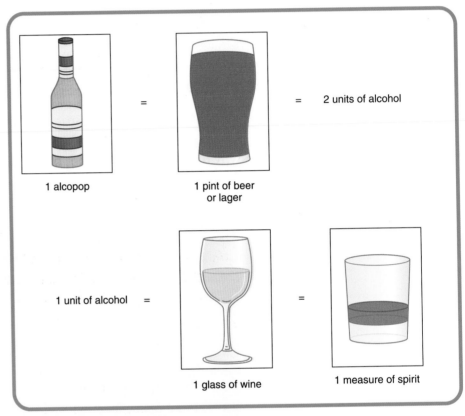

1 alcopop = 1 pint of beer or lager = 2 units of alcohol

1 unit of alcohol = 1 glass of wine = 1 measure of spirit

Figure 12.2 *Some drinks and the units of alcohol they contain*

Alcohol is broken down by the body in the liver. The rate at which this takes place is about one unit per hour. Think of a person who has eight units of alcohol in their body at midnight. It will take until about eight o'clock in the morning before all of the alcohol has been broken down.

Section Questions

1 Name two parts of the body that can be damaged by alcohol.

2 How many units of alcohol in total are present in three pints of lager and two pub measures of whisky?

3 At 11 a.m. a person has six units of alcohol in their body. At about what time would all of this alcohol have been broken down?

Making alcoholic drinks

All alcoholic drinks are made from carbohydrates present in fruits or vegetables. The carbohydrates can be either sugars or starch, as shown in table 12.1. The type of alcoholic drink varies depending on the plant source of the carbohydrate.

alcoholic drink	fruit or vegetable used	carbohydrate present
beer	barley	starch
lager	barley	starch
whisky	barley	starch
vodka	potatoes	starch
cider	apples	sugars
wine	grapes	sugars

Table 12.1 *Starting materials for some alcoholic drinks*

If starch is the starting material for an alcoholic drink, it is first broken down into the sugar called maltose. Adding yeast to a solution of maltose breaks it down into glucose. Enzymes catalyse both reactions, which involve reaction with water:

starch + water → maltose

maltose + water → glucose

Glucose is the sugar that is present in grapes. Yeast also acts on a solution of glucose in water in a process called **fermentation**. Enzymes in the yeast catalyse the breakdown of glucose into the alcohol that is called **ethanol**. Carbon dioxide gas is produced at the same time:

glucose → ethanol + carbon dioxide

How can we show that carbon dioxide is produced during fermentation? The answer is of course to pass the gas produced through lime water (see figure 12.3). Carbon dioxide turns lime water milky.

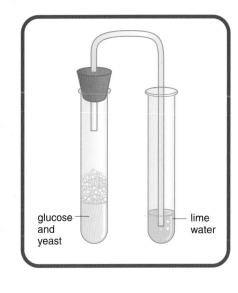

Figure 12.3 *Showing that carbon dioxide is produced during fermentation*

glucose and yeast

lime water

It is possible to make drinks with an alcohol content of up to 15% using fermentation. Some of these are shown in figure 12.4.

Figure 12.4 *These drinks are made using fermentation*

In order to obtain a drink with a higher alcohol content than 15%, the process of **distillation** must be used. During distillation, a liquid is heated until it boils and the liquid turns to a gas. The gas is then passed into a condenser, where it changes back to a liquid again. The alcohol ethanol boils at a lower temperature than water. So when a mixture of the two is heated, the ethanol distils off first. In making whisky, copper pot stills are used, like the ones in figure 12.5.

Figure 12.5 *A copper pot still used in the whisky-making industry*

Copper pot stills do not separate alcohol from water completely. They do not have to. This is because the alcohol content of whisky is usually about 40%. The rest is mainly water. Many alcoholic drinks, which are called **spirits** (see figure 12.6), also have an alcohol content of about 40%. All of them are made using both fermentation and distillation.

Figure 12.6 *All of these spirits were made using fermentation followed by distillation*

The alcohol content of drinks varies widely. Some typical values are shown in table 12.2.

drink	alcohol %
beer	5
lager	5
cider	8
wine	12
whisky	40
gin	40
vodka	40

Table 12.2 *Alcohol content of some drinks*

Ethanol and addiction

Some people find that they cannot manage without ethanol. They are then said to be addicted to it. Being unable to manage without any drug is called **addiction**.

Figure 12.7 *Methylated spirit is a very poisonous liquid*

Methylated spirit

Methylated spirit or 'meths' (see figure 12.7) is mainly made up of ethanol, but it has three other substances added to it. A purple dye lets us easily recognise it. A foul-tasting compound makes it taste awful. However, the most dangerous compound added is the alcohol **methanol**.

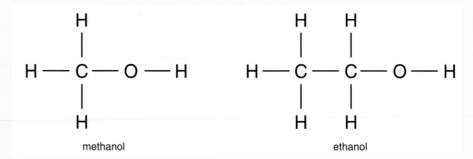

methanol   ethanol

They look similar, but methanol is much more toxic (poisonous) than ethanol. If sufficient methanol is drunk, it can cause blindness and death.

Section Questions

4 Which carbohydrate, present in barley, is the starting material for beer, lager and whisky?

5 Which gas is given off during fermentation?

6 Which process is used to increase the percentage of alcohol in drinks such as whisky, following fermentation?

7 Methylated spirit contains methanol. What effects can methanol have on a person?

Medicines

All medicines are drugs and, like alcohol, they are of course legal. If you have been injured in an accident, or if you have had an operation, very powerful drugs are used to ease pain. Morphine and codeine are two such drugs. Less powerful, but still very useful, are aspirin and paracetamol (see figure 12.8). People take aspirin or paracetamol to ease a headache, for example. All of these medicines are made to very high standards of purity. This makes their action on the body as predictable as possible.

Figure 12.8 *These common medicines alter how our bodies work in a good way*

The uses of some common medicines are summarised in table 12.3.

medicine	use
aspirin	pain relief, inflammation reduction
ibuprofen	pain relief, inflammation reduction
paracetamol	pain relief, fever reduction
codeine	pain relief, cough suppressant
milk of magnesia	acid indigestion relief

Table 12.3 *Some common medicines and their uses*

Medicines that bring pain relief are called **analgesics**. Milk of magnesia and other medicines that relieve acid indigestion are called **antacids**. They work by neutralising the excess acid. Infections are fought using **antibiotics**. These destroy micro-organisms, which can harm the body. An example of an antibiotic is amoxicillin, which is often used to treat throat infections. **Tranquillisers** can be used to treat the symptoms of anxiety, which often leads to difficulty in sleeping. These have a sedative effect, and include diazepam and lorazepam. A summary of some types of medicines and their uses is shown in table 12.4.

type of medicine	use
analgesic	pain relief
antacid	acid indigestion relief
antibiotic	fight infections by micro-organisms
tranquillisers	reduce feelings of anxiety

Table 12.4 *Some types of medicines*

How medicines work

Chemical reactions are going on in the body all the time to keep it working properly. Medicines contain drugs that help the body when it is not working properly. For example, if the stomach has produced too much acid, then an antacid (see figure 12.9) is used to neutralise the excess acid.

Micro-organisms called bacteria can cause serious infections in the body. The bacteria can multiply rapidly, destroying tissue and releasing toxins into the body. These can be spread by the blood to vital organs such as the heart and kidneys. It is fortunate that modern antibiotics can destroy these bacteria. However, some bacteria are becoming very resistant to even the most powerful antibiotics.

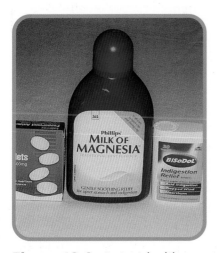

Figure 12.9 *Antacids, like these, neutralise excess acid produced by the stomach*

Medicines as mixtures

Have a look at the ingredients in an aspirin tablet (see figure 12.10 and table 12.5). In addition to aspirin, which is the 'active ingredient', there is also lactose and starch, which are inactive. Lactose improves the taste of the tablet, and starch helps hold it together.

Figure 12.10 *What is in an aspirin tablet?*

ingredients	
aspirin (300 mg)	active ingredient
lactose	inactive ingredient
starch	inactive ingredient

Table 12.5 *Active and inactive ingredients in an aspirin tablet*

Medicines are often made up of many chemicals. Some are active ingredients, and others are inactive ingredients. It is only the active ingredients that work on the body.

The active ingredient or ingredients are clearly listed on the label, along with their masses or percentages. This is true of all medicines. For example, Lemsip is used to treat the symptoms of colds and flu. Each Lemsip sachet contains the ingredients shown in table 12.6.

active ingredients	
paracetamol 650 mg	phenylephrine 10 mg
inactive ingredients	
ascorbic acid	sodium citrate
aspartame	lemon flavour
caster sugar	sodium saccharin
pulverised sucrose	curcumin
citric acid	

Table 12.6 *Active and inactive ingredients in a Lemsip sachet*

Harmful and illegal drugs

Whereas medicines and alcohol are legal drugs, there are many which are illegal. For the most part, these are not made to a high degree of purity. As a result, the way in which they behave in the body is much less certain. It is quite common for death to result from taking illegal drugs.

Some common, non-medicinal, drugs that are legal but may be harmful are shown in Table 12.7.

drug	where found	effect
alcohol	alcoholic drinks	sedative
nicotine	cigarettes	stimulant
caffeine	coffee, tea, cola	stimulant

Table 12.7 *These drugs are legal, but may be harmful*

The drugs shown in table 12.8 are all illegal.

drug	effect
amphetamines	stimulant
cocaine	stimulant
LSD (lysergic acid)	hallucinogen
cannabis (marijuana)	depressant and hallucinogen
ecstasy	stimulant and hallucinogen

Table 12.8 *All of these drugs are illegal*

Section Questions

8 Alcohol, cannabis, ecstasy, caffeine, nicotine and LSD are all drugs. Present this information in a table with the heading 'legal drugs' and 'illegal drugs'.

9 Aspirin and paracetamol are analgesics. What does this mean?

10 Taking some drugs can lead to addiction. What does this mean?

11 Ingredients in medicines are often listed as 'active ingredients' and 'inactive ingredients'. Which ones work on the body?

ACCESS 3 Subsection Test: Drugs

Part A

This part of the paper consists of four questions and is worth 4 marks.

1 Which of the following is alcohol most likely to harm? (1)

the bowel or **the brain**

2 Which of the following can alcohol be made from? (1)

starch and sugars or **starch only**

3 Spirits, like whisky, are made by (1)

fermentation and distillation or **fermentation only**

4 Which of the following is a legal drug? (1)

    caffeine or **cannabis**

Part B

This part of the paper is worth 6 marks.

5 A mixture of alcohol and water was distilled and the alcohol content of the product was noted every 5 minutes.

time / minutes	alcohol content
5	90
10	80
15	70
20	60

  a) What is the alcohol content of the product after 15 minutes of distilling? (1)

  b) Explain why it is possible to partially separate water and alcohol by distillation? (1)

6 The alcohol content of drinks varies considerably. The labels on three bottles gave these results:

  lager had an alcohol content of 6%

  wine had an alcohol content of 12%

  gin had an alcohol content of 40%

  Present this information as a table with two headings. (1)

7 Alcohol can be measured in units:

bottle of alcopop	2 units
glass of wine	1 unit

The body can break down alcohol at a rate of 1 unit per hour.

At a party a woman drinks three alcopops and two glasses of wine.

  a) How many units of alcohol did the woman consume? (1)

  b) How many hours would it be before all the alcohol had been broken down? (1)

8 Some tablets prescribed by a doctor contained the following ingredients.

active ingredient
ferrous sulphate

inactive ingredients
lactose
stearic acid
magnesium stearate

Which of the ingredients works on the body in an effort to restore the normal working of it? (1)

                                    Total 10 marks

Intermediate 1 Unit Test: Chemistry and Life

Part A

This part consists of twelve questions and is worth 12 marks.

In questions 1 to 6 choose the correct word to complete the sentences.

  1 During photosynthesis, carbon dioxide is absorbed through the **roots/leaves** of plants. (1)

2 Animals obtain energy by the reaction of glucose with **oxygen/nitrogen** to produce water and carbon dioxide. (1)

3 Herbicides save plants because they kill **fungi/weeds**. (1)

4 To be effective, fertilisers must be **insoluble/soluble** in water. (1)

5 Amino acids are examples of **monomers/polymers**. (1)

6 The fermentation of sugars produces the alcohol **ethanol/methanol**. (1)

Questions 7 to 12 are multiple choice questions. Choose the correct letter.

7 Which of the following contains nitrogen? (1)
 A fats
 B oils
 C proteins
 D carbohydrates

8 Which of the following are *not* major fertiliser compounds? (1)
 A ammonium compounds
 B copper compounds
 C nitrate compounds
 D phosphate compounds

9 Which of the following is an example of a polymer? (1)
 A starch
 B sucrose
 C glucose
 D maltose

10 Which of the following is likely to reduce the level of carbon dioxide in the atmosphere? (1)
 A increased use of aeroplanes
 B increased use of coal fires
 C increased use of cars
 D increased planting of trees

11 Which of the following elements is supplied by minerals for the healthy growth of bones and teeth? (1)
 A carbon
 B chlorine
 C hydrogen
 D calcium

12 Which of the following is believed to be true of the fats called 'saturates'? (1)

	level of cholesterol	risk of heart disease
A	lowered	lowered
B	lowered	increased
C	increased	lowered
D	increased	increased

Part B

This part consists of seven questions and is worth 18 marks.

13 The drug paracetamol is present in several medicines. One teaspoonful of Calpol contains 120 mg of paracetamol. One paracetamol tablet contains 500 mg. One Lemsip sachet contains 650 mg of paracetamol.

a) Information relating to the three medicines is presented in the table below.

medicine	amount	mass of paracetamol / mg
....	one teaspoonful	120
paracetamol tablet	one tablet	......
....	....	650

Copy and complete the table by adding the missing information. (1)

b) Lemsip contains two active ingredients and some inactive ingredients to sweeten and flavour the medicine.

active ingredients	action
paracetamol	pain relief
phenylephrine	decongestant

inactive ingredients	effect
sugar	sweetener
saccharin	sweetener
aspartame	sweetener
citric acid	flavour

i) Which **type** of ingredient is included in the medicine for the purpose of acting on the body because it is not working properly? (1)

ii) What is the action of the ingredient phenylephrine? (1)

14 A lamp was placed near to a piece of pondweed in a test tube containing water as shown.

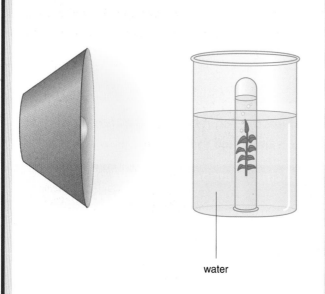

water

The number of bubbles produced by the pondweed in one minute was counted, and the distance of the lamp from the test tube was measured. The lamp was moved to a new position and the measurements were repeated. This was done several times.

The following graph was produced.

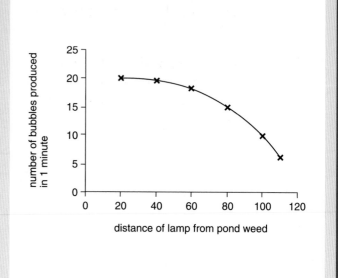

a) How many bubbles of gas were produced in one minute when the lamp was 30 cm from the pondweed? (1)

b) Predict how many bubbles of gas would be produced in one minute when the lamp was 120 cm from the pondweed. (1)

c) The gas produced relit a glowing splint. Which gas gives this result? (1)

15 A sample of starch solution was tested with Benedict's solution, but no change in colour took place.

The enzyme amylase was then added to the starch solution with the temperature maintained at about 37°C. After a few minutes a sample of the solution was tested with Benedict's solution and gave a positive result.

a) Name a sugar that the starch could have been broken down into by the action of the enzyme. (1)

One of the pupils mistakenly boiled the mixture containing the amylase and starch. He then found that the Benedict's test did not work.

b) Explain why amylase, a body enzyme, would not be able to catalyse the breakdown of starch once it had been heated strongly. (1)

16 A police force investigated the link between road traffic accidents and drinking while over the legal limit. Over a period of years they compiled the following statistics.

year	total number of road traffic accidents	number of accidents linked to drink driving
2000	250	130
2001	230	135
2002	210	140

a) Describe the trend in the total number of road traffic accidents in the period from 2000 to 2002. (1)

b) What is the average number of accidents that were linked to drink driving over the three years? (1)

17 Vitamins and minerals are often added to our food. These are then listed on the packaging. For example, margarine has vitamins A and D added to it.

a) Why are vitamins added to food? (1)

One particular brand of margarine gave the following information on its label.

nutritional information	(per 100 g of product)
total fats	35 g
of which polyunsaturates	17.5 g

b) What is the percentage of polyunsaturates in the total fat content of the margarine? (1)

Compared to many other margarines, the one referred to above has a high proportion of polyunsaturates.

c) Why should people decrease the amount of saturates in their diet and increase the amount of polyunsaturates they consume? (1)

18 Alcohol contents of drinks vary widely. Some concentrations are shown below.

lager	5%
wine	15%
whisky	40%

a) Present this information as a bar graph. (1)

b) Name two organs of the body that are very likely to be damaged by the excessive consumption of alcohol. (1)

19 In an experiment investigating the behaviour of carbon dioxide, the apparatus shown below was used.

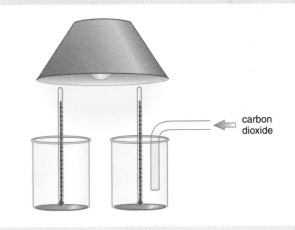

carbon dioxide

The following readings of the thermometers in the two beakers were noted every 2 minutes.

beaker A		beaker B	
time / minutes	temperature / °C	time / minutes	temperature / °C
0	20	0	20
2	22	2	23
4	24	4	26
6	26	6	29
8	28	8	32

a) Plot two line graphs of these results. (2)

b) Carbon dioxide levels in the atmosphere are rising. What effect on climate is this believed to be causing? (1)

Total 30 marks

Unit 3 Glossary of Terms

addiction   Being unable to manage without a drug.

amino acids   These are formed when proteins are broken down.

antibiotics   Drugs that fight and destroy micro-organisms, which interfere with chemical reactions in the body.

Benedict's solution   Used to test for the presence of the sugars glucose, fructose or maltose. The colour change is from blue to orange-red.

carbohydrates   Compounds containing carbon, hydrogen and oxygen that supply the body with energy.

chlorophyll   A green compound in the leaves of plants. It absorbs the light energy needed for photosynthesis.

digestion   The breaking down of large food molecules into smaller ones.

distillation   This is a method of increasing alcohol concentration of fermentation products. This happens because alcohol (ethanol) has a lower boiling point than water.

drugs   Substances that alter how the body works.

ethanol   The chemical name for alcohol. Taken in excess, this can cause damage to the brain and liver.

fats   Together with oils, they form a concentrated energy source for our diet. Both are compounds of carbon, hydrogen and oxygen.

fermentation   This is the breakdown of glucose into alcohol and carbon dioxide, catalysed by enzymes in yeast. Wines, beers, lagers and cider are made by fermentation.

fertiliser   A natural or artificial substance that is added to soil to restore essential elements.

fibre   Important in our diet, as it keeps the gut working well, preventing constipation.

food additives   These include vitamins, minerals, preservatives, colourings and flavourings.

fungicides   Chemicals that prevent diseases in plants by killing fungi and bacteria.

greenhouse effect   This is caused by carbon dioxide and some other gases. They cause the atmosphere to retain more of the sun's heat energy. This results in global warming.

herbicides   Chemicals that kill weeds.

illegal drugs   These include cannabis, LSD and ecstacy.

iodine solution   Used to test for starch. The colour change is from reddish-brown to dark blue.

legal drugs   These include alcohol, nicotine and caffeine.

medicines   Contain drugs that help the body when it is not working properly. Only the active ingredients in a medicine work on the body.

methanol   This is a very toxic alcohol. It can cause blindness and death.

methylated spirit   A liquid that is mainly ethanol, but to which has been added methanol, a purple colour and a bad tasting substance.

minerals   Compounds in the diet that supply the body with calcium for bones and teeth, iron for the blood, as well as trace elements.

pesticides   Toxic chemicals used to control pests.

photosynthesis   A process during which plants use light energy to convert carbon dioxide and water into glucose and oxygen. This happens in plant leaves.

polyunsaturates　These are polyunsaturated fats and oils, which are considered to be less harmful to the heart than saturates.

predator　An animal that hunts another animal for its food.

proteins　These important food polymers provide material for body growth and repair. During digestion they are broken down to amino acids. All contain carbon, hydrogen, oxygen and nitrogen.

respiration　A process during which animals and plants obtain energy from glucose. Glucose reacts with oxygen to produce carbon dioxide, water and energy.

root nodules　Lumps on the roots of peas, beans and clover in which nitrogen from the air is converted into nitrates.

saturates　These are saturated fats. They are believed to increase the level of cholesterol in the blood, which may cause heart disease.

starch　A very large polymer carbohydrate. It is not sweet and does not dissolve readily in water. It is an energy store in plants.

sugars　Small carbohydrate molecules that are soluble and sweet tasting, e.g. glucose, fructose, maltose and sucrose.

toxic　Poisonous.

unit of alcohol　This is equal to half a pint of beer, a pub measure of spirit or a glass of wine. Alcohol is broken down in the body at about one unit per hour.

vitamins　Complex carbon compounds that are required to keep the body healthy.

Whole-Course Material

In this unit you will find
Whole-course questions
Chemical dictionary

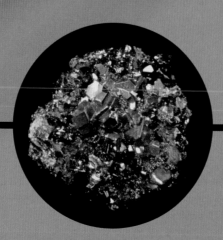

13 Whole-course questions

Section A

1 The structure of substances can be represented by models.

Which model shows an element?

A

B

C

D

2 Which hazard label would be used to indicate that a weedkiller is toxic?

A   B   C   D

3 You may wish to use page 6 of the data booklet to answer this question.

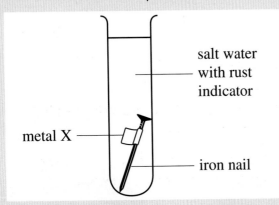

salt water with rust indicator

metal X

iron nail

In the above experiment, a blue colour appeared in the salt water.

Metal X could have been

A aluminium

B magnesium

C tin

D zinc.

4 Which amount of drink would the body break down in the shortest time?

A 2 glasses of wine

B 1 whisky

C 1 bottle of alcopop

D 1 pint of beer

5 Anodising increases the thickness of the oxide layer of

A aluminium

B iron

C magnesium

D zinc.

6 Mercury and bromine are both elements which are

A metals

B non-metals

C gases at room temperature

D liquids at room temperature.

7 The diagram shows two types of bonds in water, bonds between atoms in the molecules and bonds between molecules.

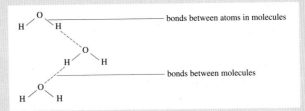

Which line in the table correctly shows how strong these bonds are?

	Bonds between atoms in molecules	Bonds between molecules
A	weak	weak
B	strong	strong
C	strong	weak
D	weak	strong

8 Which process is represented by the following word equation?

glucose + oxygen → carbon dioxide + water

A Respiration
B Fermentation
C Polymerisation
D Photosynthesis

9

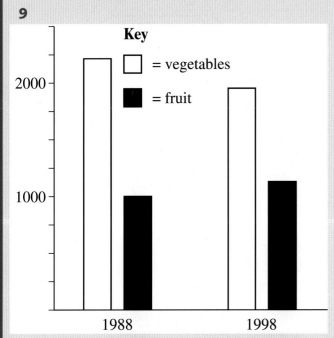

Compare with 1988, in 1998 people ate
A more vegetables and less fruit
B more vegetables and more fruit
C less vegetables and less fruit
D less vegetables and more fruit.

10 Ethanol burns to produce carbon dioxide.

Which apparatus would be used to show that carbon dioxide is formed when ethanol burns?

A

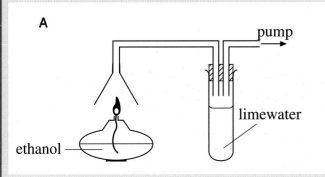

B

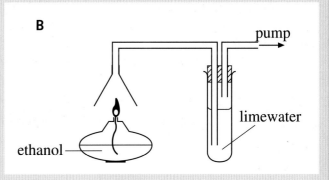

C

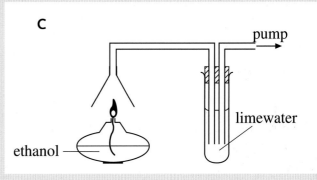

D

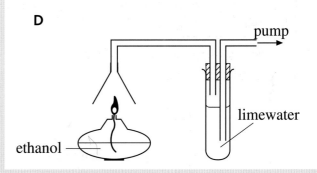

11 Which element shows similar chemical properties to chlorine?

(You may wish to use page 1 of the data booklet to answer this question.)

A Argon
B Iodine
C Oxygen
D Sulphur

12 In which experiment will the reaction be fastest?

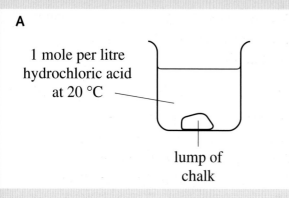

A

1 mole per litre
hydrochloric acid
at 20 °C

lump of
chalk

B

1 mole per litre
hydrochloric acid
at 30 °C

lump of
chalk

C

1 mole per litre
hydrochloric acid
at 20 °C

powdered
chalk

D

1 mole per litre
hydrochloric acid
at 30 °C

powdered
chalk

13 The fire triangle tells us that a fire needs a fuel, oxygen and a temperature high enough to start the fire and keep it going.

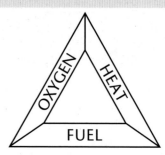

Using a fire blanket puts out fires by
A soaking up the fuel
B stopping oxygen getting to the fuel
C lowering the temperature of the fuel
D providing carbon dioxide to put out the fire.

14 Which of the following polymers is **not** a plastic?
A Bakelite
B Kevlar
C Silicone
D Starch

15 Which statement about methanol is **false**?
A It is very toxic.
B It is an alcohol.
C It can cause blindness and death.
D It is used to make alcoholic drinks.

16 Helium, neon and argon are in the same column of the Periodic Table because they are
A non-metals
B found in air
C gases at room temperature
D elements with similar chemical properties.

17 Which line in the table correctly describes what happens if 1 gram of a catalyst is involved in a chemical reaction?

	Speed of reaction	Mass of catalyst left at end in grams
A	unchanged	0
B	faster	0
C	unchanged	1
D	faster	1

18 The boiling point of a hydrocarbon depends on the size of the hydrocarbon molecule. Which of the following hydrocarbons has the lowest boiling point?

A C_6H_{14}
B C_8H_{18}
C $C_{10}H_{22}$
D $C_{12}H_{26}$

19 What percentage of body weight is water?
A Less than 30%
B Approximately 40%
C Approximately 50%
D More than 60%

20 Alcohol is made by the fermentation of glucose.

Distillation can then be used to
A increase the alcohol concentration
B decrease the alcohol concentration
C increase the glucose concentration
D decrease the glucose concentration.

Section B

1 Chemists use symbols and formulae to represent atoms, ions and molecules. Some of these are shown below.

CH_4 Na^+ Br_2 Ne Cl^- Mg

Complete the table by putting the symbols and formulae in the correct column.

Atoms	Ions	Molecules

(2)

2 Problems can be caused if poisonous liquids seep from rubbish tips into water supplies. A rubbish tip in the North of Scotland is being lined with poly(ethene) strips to prevent this. Heat is used to join the edges of the poly(ethene) strips together.

a) Name the monomer that is used to make poly(ethene). 1

b) Name the type of reaction which is used to make poly(ethene). 1

c) Suggest a reason why a biodegradeable plastic **should not** be used to line the rubbish tip. 1

d) What name is given to plastics that can be heated and reshaped? 1

(4)

3 There are two methods of producing aluminium. One is by smelting its ore and the other is by recycling old aluminium. The table shows the mass of various pollutants produced by each method per tonne of aluminium.

Pollutant	Mass of pollutant produced per tonne of aluminium	
	From its ore (in kilograms)	By recycling (in kilograms)
sulphur dioxide	89·0	1·0
dust	7·0	1·5
carbon monoxide	35·0	2·5
nitrogen oxides	139·0	7·0
hydrocarbons	87·0	5·0

a) **From the table** suggest an advantage of recycling. 1

b) Draw a bar chart to show the different pollutants produced **by recycling**. 2

c) What pollution problem is caused if sulphur dioxide is released into the air? 1

(4)

4   The following information is given on the label of a bottle of cough medicine.

a) What is an **active ingredient**?   1

b) Each dose of medicine weighs 10 g and contains 0·5g of lemon juice.

Calculate the percentage of lemon juice in the medicine.   1

c) Throat infections are caused by micro-organisms.

What type of drug might a doctor prescribe for a throat infection?   1
                                                                    (3)

5   Some street lights glow red when they are warming up.

Once they are warm they give a bright orange light.

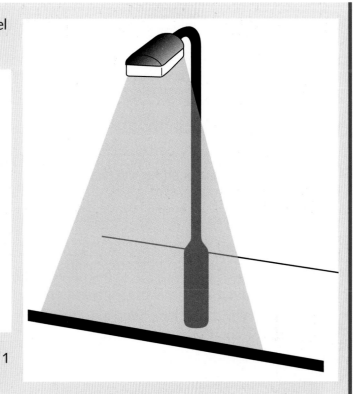

(You may wish to refer to your data booklet to answer the following questions.)

a) The red colour is caused by the gas neon.

   When was neon discovered?   1

b) The orange colour is caused by the element sodium.

   Write the symbol for the element sodium.   1
                                                        (2)

6   The diagram represents a molecule of **butane**.

$$H-\overset{\overset{\displaystyle H}{|}}{\underset{\underset{\displaystyle H}{|}}{C}}-\overset{\overset{\displaystyle H}{|}}{\underset{\underset{\displaystyle H}{|}}{C}}-\overset{\overset{\displaystyle H}{|}}{\underset{\underset{\displaystyle H}{|}}{C}}-\overset{\overset{\displaystyle H}{|}}{\underset{\underset{\displaystyle H}{|}}{C}}-H$$

a) Write the formula for butane.   1

b) Butane contains only carbon and hydrogen.

What name is used to describe compounds containing only carbon and hydrogen? 1

c) When butane burns it combines with oxygen to produce carbon dioxide and water.

Write a word equation for the reaction taking place. 1

(3)

7 A chemistry class was investigating the breakdown of starch using the enzyme amylase.

They carried out experiments at different pH values and timed how long it took for the starch to be completely broken down.

pH	Time taken for starch to break down (mins)
5·0	6·0
5·5	5·5
6·0	4·0
6·5	1·8
7·0	1·5
7·5	2·5
8·0	3·5

a) Plot these results as a line graph using the axes provided; one axis has been labelled and scaled for you. 2

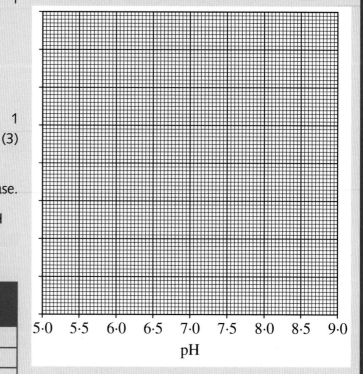

b) Predict how many minutes it will take for the starch to break down when the pH is 8·5. 1

c) Does the enzyme amylase work best in acidic, alkaline or neutral conditions? 1

d) The enzyme amylase is a natural catalyst.

What is a catalyst? 1

(5)

Oxide	Type of oxide	Effect on damp pH paper
sulphur dioxide	non-metal	turns red
sodium oxide	metal	turns blue
carbon dioxide	non-metal	turns red
calcium oxide	metal	turns blue

a) From the table, name an oxide which dissolves in water producing an alkaline solution. 1

b) Predict the effect lithium oxide would have on damp pH paper. 1

(You may wish to use page 1 of the data booklet to answer this question.)

(2)

8 a) Why do farmers spread fertilisers on fields after harvesting crops? 1

b) Peas and clover are plants which have root nodules.

What do these enable the plants to do? 1

c) A 20 kg bag of fertiliser contains 7 kg of nitrogen.

What is the percentage of nitrogen in the fertiliser? 1

(3)

9 The table gives information about some oxides.

10 a) Name the elements present in sodium carbonate. 1

b) The diagram shows that when calcium chloride solution and sodium carbonate solution are mixed a chemical reaction takes place.

i) What evidence is there that a chemical reaction has taken place? 1

ii) **Draw** and **label** a diagram of the apparatus which would be used to separate the precipitate from the solution.

Show on the diagram where the precipitate would collect. 2

(4)

calcium chloride solution   +   sodium carbonate solution   →   solution / precipitate

11 Vitamins are needed by the body to keep it healthy.

Vitamin C and Vitamin E are both found **in green vegetables**.

Vitamin A is **needed to fight disease**, while the body needs **Vitamin D** to help our bones develop properly.

Use this information to complete the key below. (2)

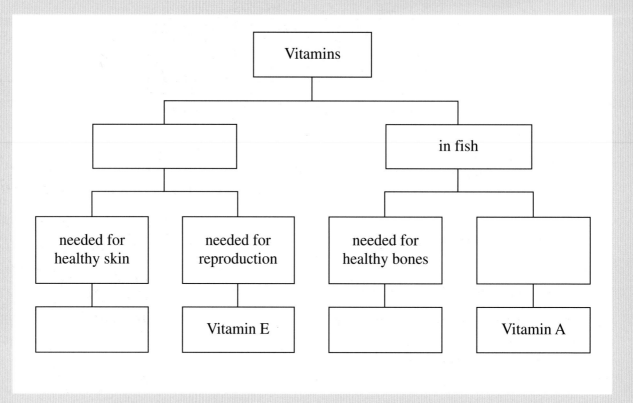

12 a) Farmers use herbicides to control weeds. The diagram shows a molecule of a herbicide.

$$
\begin{array}{c}
\quad\;\; O \qquad\;\; H \qquad\;\; H \;\; O \\
\quad\;\; \| \qquad\;\;\; | \qquad\;\;\; | \;\;\; \| \\
H-O-P \;\!\rule{1.2cm}{0.4pt}\!\; C-N-C-C-O-H \\
\quad\;\; | \qquad\qquad\;\; | \;\;\; | \;\;\; | \\
\quad\;\; O-H \quad\; H \;\; H \;\; H
\end{array}
$$

Complete the formula to show the number of each type of atom in this molecule. 1

    C H O N P

b) Why do crops not grow as well when weeds are present? 1

c) What type of compound can farmers use to prevent diseases in plants? 1

(3)

13 The table shows the chemical names of four ores.

Ore	Chemical name
bauxite	aluminium oxide
galena	lead sulphide
malachite	copper carbonate
sphalerite	zinc sulphide

The graph shows how much metal is produced each year from these ores.

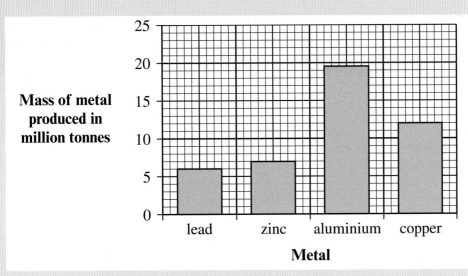

Mass of metal produced in million tonnes (y-axis, scale 0 to 25)

Metals (x-axis): lead, zinc, aluminium, copper

Metal

a) Which ore is a compound made up of **three** elements? 1

b) How much metal is produced each year from the ore galena in million tonnes? 1

(2)

14 a) Why must our diet contain proteins? 1

b) Carbon, hydrogen and oxygen atoms are found in proteins.
What other atom is always found in proteins? 1

c) When a protein is heated with soda lime a gas is given off.
This gas is detected using damp pH paper.

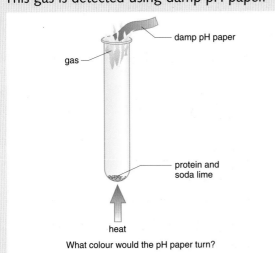

damp pH paper

gas

protein and soda lime

heat

What colour would the pH paper turn?

What colour would the pH paper turn? 1

(3)

15 The pie chart shows the percentage of phosphoric acid used to make other substances.

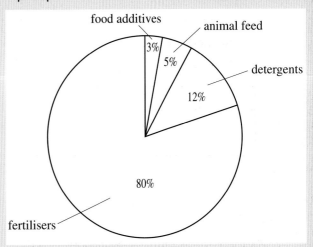

food additives 3%
animal feed 5%
detergents 12%
fertilisers 80%

a) Give **one** reason why a food might contain a food additive. 1

b) 50 million tonnes of phosphoric acid are produced in the world each year.

Calculate the mass of phosphoric acid used to make animal feed each year in million tonnes. 1

c) How does using detergent help to remove grease when washing dishes? 1

d) Name the element in calcium phosphate fertiliser which is essential for healthy plant growth. 1

(4)

Chemical dictionary

A

acid   A substance that gives a solution with a pH of less than 7.

acid rain   Rain that is more acidic than usual due to dissolved sulphur dioxide and nitrogen dioxide.

addiction   Being unable to manage without a drug.

air   A mixture of gases, approximately 80% nitrogen and 20% oxygen.

alkali   A substance that gives a solution with a pH of more than 7.

alkali metals   Reactive metal elements in column 1 of the Periodic Table.

alloy   A substance that is a mixture of metals, or of metals and non-metals. Examples include brass, solder and stainless steel.

amino acids   These are formed when proteins are broken down.

anodising   A process that increases the thickness of the oxide layer on aluminium (to provide protection against corrosion).

antibiotics   Drugs that fight and destroy micro-organisms, which interfere with chemical reactions in the body.

atom   The smallest part of an element that can exist.

atomic number   A special number given to each element in the Periodic Table.

B

battery   Batteries produce electricity from chemical reactions. They convert chemical energy into electrical energy.

Benedict's solution   Used to test for the presence of the sugars glucose, fructose or maltose. The colour change is from blue to orange-red.

biodegradable   Able to be broken down by bacteria in the soil and rot away.

biogas   A mixture of gases formed by the decomposition of plant or animal material. Consists mainly of methane.

Biopol   A biodegradable plastic.

burning   A chemical reaction during which a substance combines with oxygen, producing heat and light.

C

carbohydrates   Compounds containing carbon, hydrogen and oxygen that supply the body with energy.

catalyst   A substance that speeds up a reaction but is not used up by the reaction.

chemical reaction   A chemical process in which one or more new substances are formed (usually accompanied by an energy change and a change in appearance).

chemical symbol   One or two letters used to represent an element, for example C for carbon, Al for aluminium.

chlorophyll   A green compound in the leaves of plants. It absorbs the light energy needed for photosynthesis.

combustion   Another word for burning.

compound   A substance in which two or more elements are joined together chemically. Compounds are formed when elements react together.

corrosion A chemical reaction that involves the surface of a metal changing from an element to a compound.

cracking An industrial method of breaking up larger hydrocarbon molecules to produce smaller, more useful molecules.

D

digestion The breaking down of large food molecules into smaller ones.

distillation A process of separation based on differences in boiling points. The changes of state involved are:

liquid → gas → liquid.

It is a method of increasing alcohol concentration of fermentation products. This happens because alcohol (ethanol) has a lower boiling point than water.

drugs Substances that alter how the body works.

dry-cleaning A process using special solvents that are particularly good at dissolving oil and grease stains.

durable Long lasting, hard wearing.

dyes Coloured compounds that are used to give bright colours to clothing.

E

electroplating A process by means of which a layer of metal is deposited on a substance using electricity. The object is used as the negative electrode in a solution containing ions of the metal being deposited.

element The simplest kind of substance that cannot be broken down into anything simpler.

enzyme A catalyst that affects living things.

ethanol The chemical name for alcohol. Taken in excess, this can cause damage to the brain and liver.

F

fats Together with oils, they form a concentrated energy source for our diet. Both are compounds of carbon, hydrogen and oxygen.

fermentation This is the breakdown of glucose into alcohol and carbon dioxide, catalysed by enzymes in yeast. Wines, beers, lagers and cider are made by fermentation.

fertiliser A natural or artificial substance that is added to soil to restore essential elements.

fibre Important in our diet, as it keeps the gut working well, preventing constipation.

fibres Thin strands used to make, among other things, clothing fabrics.

finite resources Ones that there are only limited supplies of and that cannot be replaced.

fire triangle A way of understanding how fires can be put out. This can be done by removing heat or fuel. The fire will also go out if oxygen cannot reach it.

food additives These include vitamins, minerals, preservatives, colourings and flavourings.

fossil fuels Fuels formed from plant and animal material over a very long time. Examples include coal, oil, natural gas and peat.

fraction A group of compounds with boiling points within a given range. Petrol, kerosene and diesel are examples of fractions.

fractional distillation The process used to separate crude oil into fractions according to the boiling points of the components of the fractions.

fuel A substance that is burned to produce heat energy.

fuel crisis Having not enough supplies of fuel to meet the demand for it.

fungicides   Chemicals that prevent diseases in plants by killing fungi and bacteria.

G

galvanising   A process by which iron is coated with a protective layer of zinc (by dipping into molten zinc).

greenhouse effect   This is caused by carbon dioxide and some other gases. They cause the atmosphere to retain more of the sun's heat energy. This results in global warming.

H

halogens   Reactive non-metal elements in column 7 of the Periodic Table.

herbicides   Chemicals that kill weeds.

hydrocarbon   A compound that contains hydrogen and carbon only. An example is methane, CH_4.

I

illegal drugs   These include cannabis, LSD and ecstacy.

incinerate   A process of burning refuse, such as waste plastic, so that it is reduced to ashes.

iodine solution   Used to test for starch. The colour change is from reddish-brown to dark blue.

ion   An atom or group of atoms that possess a positive or negative charge.

L

legal drugs   These include alcohol, nicotine and caffeine.

M

malleability   The ability to be shaped by hammering, rolling and bending. A physical property of metals.

medicines   Contain drugs that help the body when it is not working properly. Only the active ingredients in a medicine work on the body.

metals   Shiny elements that conduct electricity.

methanol   This is a very toxic alcohol. It can cause blindness and death.

methylated spirit   A liquid that is mainly ethanol, but to which has been added methanol, a purple colour and a bad tasting substance.

minerals   Compounds in the diet that supply the body with calcium for bones and teeth, iron for the blood, as well as trace elements.

mixture   A mixture is formed when two or more substances come together but do not react, e.g. air and sea water.

molecule   A group of two or more atoms held together by strong bonds.

monomer   A small molecule from which large polymer molecules can be made.

N

natural fibres   Ones that come from plants and animals. Examples include wool, silk and cotton.

neutral solution   One with a pH of 7.

neutralisation reaction   A reaction of acids with alkalis or metal carbonates that moves the pH of the solution towards 7.

noble gases   Unreactive gases in column 0 of the Periodic Table.

non-metals   Elements that do not conduct electricity (carbon is an exception). Mostly not shiny.

P

Periodic Table   An arrangement of the elements in order of increasing atomic number. Chemically similar elements occur in the same vertical column.

pesticides   Toxic chemicals used to control pests.

pH   A number that indicates the degree of acidity or alkalinity of a solution. The pH scale ranges from below 0 (very acidic) to above 14 (very alkaline).

photosynthesis   A process during which plants use light energy to convert carbon dioxide and water into glucose and oxygen. This happens in plant leaves.

plastics   Synthetic materials made up of polymers.

pollutant   Something that harms the environment.

polymer   A very large molecule that is formed by the joining together of many small molecules called monomers.

polymerisation   The process of joining together many small monomer molecules to make a polymer.

polyunsaturates   These are polyunsaturated fats and oils, which are considered to be less harmful to the heart than saturates.

predator   An animal that hunts another animal for its food.

proteins   These important food polymers provide material for body growth and repair. During digestion they are broken down to amino acids. All contain carbon, hydrogen, oxygen and nitrogen.

R

recycle   Convert waste to re-usable material. For example, waste polythene shopping bags are converted into bin bags.

renewable resource   Resources that can be replaced. Examples of renewable sources of energy include methane, ethanol and hydrogen.

respiration   A process during which animals and plants obtain energy from glucose. Glucose reacts with oxygen to produce carbon dioxide, water and energy.

root nodules   Lumps on the roots of peas, beans and clover in which nitrogen from the air is converted into nitrates.

rust indicator   A pale yellow solution, which turns blue if rusting is taking place.

rusting   The corrosion of iron. It is caused by the reaction of iron with oxygen and water.

S

salt   A compound formed by the result of a reaction between an acid and an alkali or metal carbonate, e.g. sodium chloride.

saturated solution   One in which no more substance can be dissolved.

saturates   These are saturated fats. They are believed to increase the level of cholesterol in the blood, which may cause heart disease.

scum   A solid formed by the reaction of some soaps with hard water.

silicones   Compounds used for water-proofing fabrics.

solution   A liquid with a substance dissolved in it.

starch   A very large polymer carbohydrate. It is not sweet and does not dissolve readily in water. It is an energy store in plants.

sugars   Small carbohydrate molecules that are soluble and sweet tasting, e.g. glucose, fructose, maltose and sucrose.

symbol   *See* chemical symbol.

synthetic   Man-made.

synthetic fibres   Ones that are made by the chemical industry. Examples include nylon and polyesters such as Terylene.

T

thermoplastic   Plastic that softens on heating and can be reshaped.

thermosetting plastic   Plastic that does not soften on heating and cannot be reshaped.

toxic   Poisonous.

U

unit of alcohol   This is equal to half a pint of beer, a pub measure of spirit or a glass of wine. Alcohol is broken down in the body at about one unit per hour.

V

vitamins   Complex carbon compounds that are required to keep the body healthy.

W

word equation   An equation that gives the names of reactants and products, e.g.

carbon + oxygen → carbon dioxide.

Page numbers relating to Figures, Tables or illustrations are in *italic* print. Numbers relating to practical activities are in **bold** print.